THE NEXT BIG THING

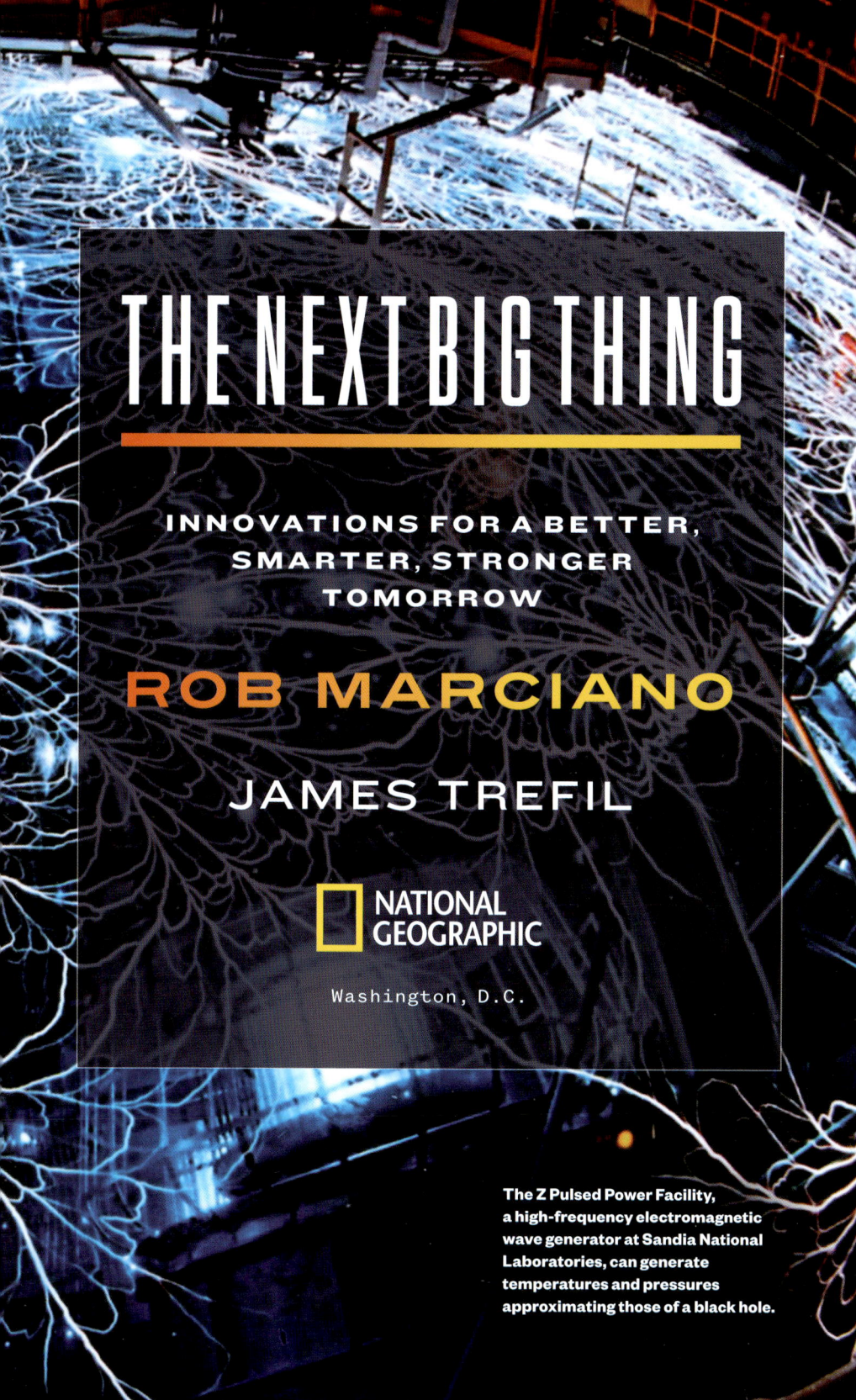

THE NEXT BIG THING

INNOVATIONS FOR A BETTER, SMARTER, STRONGER TOMORROW

ROB MARCIANO

JAMES TREFIL

NATIONAL GEOGRAPHIC

Washington, D.C.

The Z Pulsed Power Facility, a high-frequency electromagnetic wave generator at Sandia National Laboratories, can generate temperatures and pressures approximating those of a black hole.

Contents

III

INFRASTRUCTURE

IV

INFORMATION

INTRODUCTION

The Milky Way appears over a solar panel array in Poland. Solar power is among the most used renewable energy sources in the country.

A **Future** Shaped by **Science**

A AN EARDRUM-PIERCING WHISTLE fills the air as the wind screams around trees, streetlights, and signposts. Building materials, landscaping, and traffic lights thrash across the hotel parking lot with deadly speed, some of the debris completely airborne. I cower against the cement wall beside the lobby entrance, feeling the building sway, pushed to its physical limit, as I report live on CNN from Biloxi, Mississippi. It's only my second experience inside a major hurricane, and Katrina was unleashing a fury on the Gulf Coast of the United States beyond anything I could fathom.

We have a saying in the weather community: You can hide from the wind, but you must run from the water. As my team reports from Biloxi on August 29, 2005, a storm surge rises from the Gulf of Mexico just two miles (3.2 km) south of our location, and a wall of water as high as 28 feet (8.5 m) devours the heart of Biloxi, sweeping entire commercial buildings and residential homes clean from their foundations. To the west, after the eye of the hurricane has careened through, billions of gallons of water breach the levees in New Orleans and flood the city. Through my internal feedback earpiece, I hear my colleagues in New Orleans wading through

Modern computer chips include billions of transistors—essentially tiny information storage units that undergird our digital world. These chips are common today, but quantum computer chips might be the future of digital technology (see pages 262–271).

rising floodwaters, sifting through debris, and helping victims as they report on the disaster. Ultimately, Hurricane Katrina would claim nearly 1,400 lives across Louisiana and Mississippi. As of this writing, it's still the costliest natural disaster in U.S. history.

Storm-chasing meteorologists like me often view weather events through the lens of potential—but not just their potential to destroy. Hurricanes actually have some positive potential. They help regulate the planet's uneven ocean temperatures, nourish drought-stricken land, fatten rivers and streams, and prune weak or dying trees to renew forests. All tropical cyclones, including hurricanes, act to release the oceans' built-up summertime heat into the air, and often transport that tropical warmth from the Equator to the poles. So in essence, as the climate crisis warms the planet, hurricanes fight to keep things cool.

But as Katrina demonstrated, nature's interests sometimes compete with our own. Our modern conveniences have pulled us into habits that contribute to warming global temperatures. We power things with limited resources, we travel in emissions-spouting vehicles, and we produce most of our goods using cheap plastics. Make no mistake, many of our modern innovations are productive, efficient, and exciting, but they find us at a breaking point with our planet. Are we going to continue down this path, or find a way to sustainability? Like a hurricane, we have potential in two directions. We could bring ourselves further harm, or we can turn toward a better, smarter, stronger tomorrow.

It might be an uphill battle, but if I've come to believe in anything during my career as a reporter, it's humankind's ability to problem-solve, and to use science as our best tool to do so. When our news team was in the bowels of Hurricane Katrina, numerous innovations helped us report from the storm: satellite phones to communicate with CNN headquarters, mobile generators to keep our crew functioning without power from the grid, waterproof materials to maintain equipment operations in the wind and wet, and blessed mobile hotspots to connect us to the web and other news teams along the coast. Science has helped me do my job for a long time, but I think it can do so much more. In fact, I believe science holds the answers to building a prosperous, healthy outcome for both ourselves and our planet, and this book is an exploration of innovations that could tip us toward a positive future. These breakthroughs are the *next big thing*.

We will find the next big thing in four tentpole realms of science: energy, electricity, infrastructure, and information. These topics encompass nearly everything we use as humans, including what we drive, live in, work on, play with, and build. The applications can be mind-boggling—electric aircraft, 3D-printed houses, AI chatbots, quantum computers—but the science beneath is about as basic as what we learned in high school.

The world of **Energy** involves how we generate power. It might be the most important frontier for scientific innovation this century, as the hunt for green and sustainable energy could single-handedly determine how we meet the climate crisis. We'll take apart a wind turbine and survey a solar power plant for a look at current renewables, and then we'll see what's around the corner. Iceland has tapped geothermals to go green, and Hoover Dam might transform into a giant energy-saving battery. We'll hunt for some long-term energy solutions, too—solutions that produce the same limitless energy as the sun.

The **Electricity** section evokes my time chasing tornadoes for ABC News. In 2015, my team was charging after thunderstorms in southwestern Oklahoma when a lightning bolt made impact about 200 yards (180 m) from where we were standing. It scared us out of our shoes, and we ran to the safety of our vehicles, screaming like children. I often think of that incredible power when I see Teslas on the road or pass the transformer in my neighborhood. Lightning strikes are basically electric currents between the ground and clouds above, and the same principle powers the grid, our homes, and our gadgets. We'll unpack why electric current is such a clean and efficient fuel source, and then we'll use it to go for a spin in a zero-carbon electric aircraft. Sure, autonomous electric cars are cool and all, but why sit in traffic when you can take to the skies?

Infrastructure involves the systems and processes we use to build things. Concrete and steel rule the construction world, but extracting the sand we need for concrete depletes it from riverbeds and shorelines. For every metric ton (1.1 U.S. tons) of steel we produce, nearly two metric tons (2.2 U.S. tons) of carbon dioxide (CO_2) are belched into the air. Engineers are finding inventive ways to make these materials healthier for the planet—like carbon-neutral concrete that can heal itself, or special alloyed steel that lasts longer. But they're also tapping ancient materials for modern uses. The tried-and-true printing process turned three-dimensional will be a hyperefficient tool for builders of tomorrow. And, believe it or not, wood is the sustainable backbone supporting one of the most beautiful high-rise buildings I have ever seen—wait until you see it.

In my business, **Information** is everything. The more weather data you can wrangle, the more accurate the forecast it will yield. In fact, sophisticated tornado chasers now use heavy-duty vehicles called DOWs, Doppler on Wheels, to follow storms. DOWs are modified tractor trailers equipped with meteorological instrumentation, most notably a rotating Doppler dish with intricate sensors that can look inside clouds and measure not only precipitation, but also wind direction and velocity: functionally, an x-ray for tornadoes. Meteorologists of 100 years ago could only dream of having this kind of access to information. It's making a transformative impact across all industries, and all because of digitization, computers, and artificial intelligence. But this stuff can be baffling. Can we really know what goes on inside the "mind" of a machine? To find out, I talked to one of the world's most advanced robots about how she works and about artificial intelligence. Her answers illuminated a world of binary code, computer chips, and neural networks. It's all real science, rooted in surprisingly simple concepts.

My journey to find the next big thing became a journey to learn how things work—even things I take for granted, such as my smartphone and the internet. I found answers that blew my mind and answers that struck me with their approachability. Often, when we talk about the future, there are a lot of worst-case scenarios. And it's true that if we don't change the way we do nearly everything, entire ecosystems could collapse. But right now—especially after I've taken a deep dive into the innovative science and technologies on our horizon—I'm hopeful we can meet that crisis and change our planet's future. As those TV scientists said when they built the Six Million Dollar Man, "We have the technology."

With our mastery of science and engineering, the possibilities for our future are endlessly promising. We're living in one of the most challenging, yet exciting, moments in human history. Significant problems face us, and the conflict can be tense, but my curiosity about the future, shared in this book, has allowed a wealth of knowledge and perspective to seep into my soul. I'm excited for you, the readers, to experience that hopeful thrill as well. We have so many reasons to be optimistic about the next big thing. If you ask me, it's not even around the corner. It's already here!

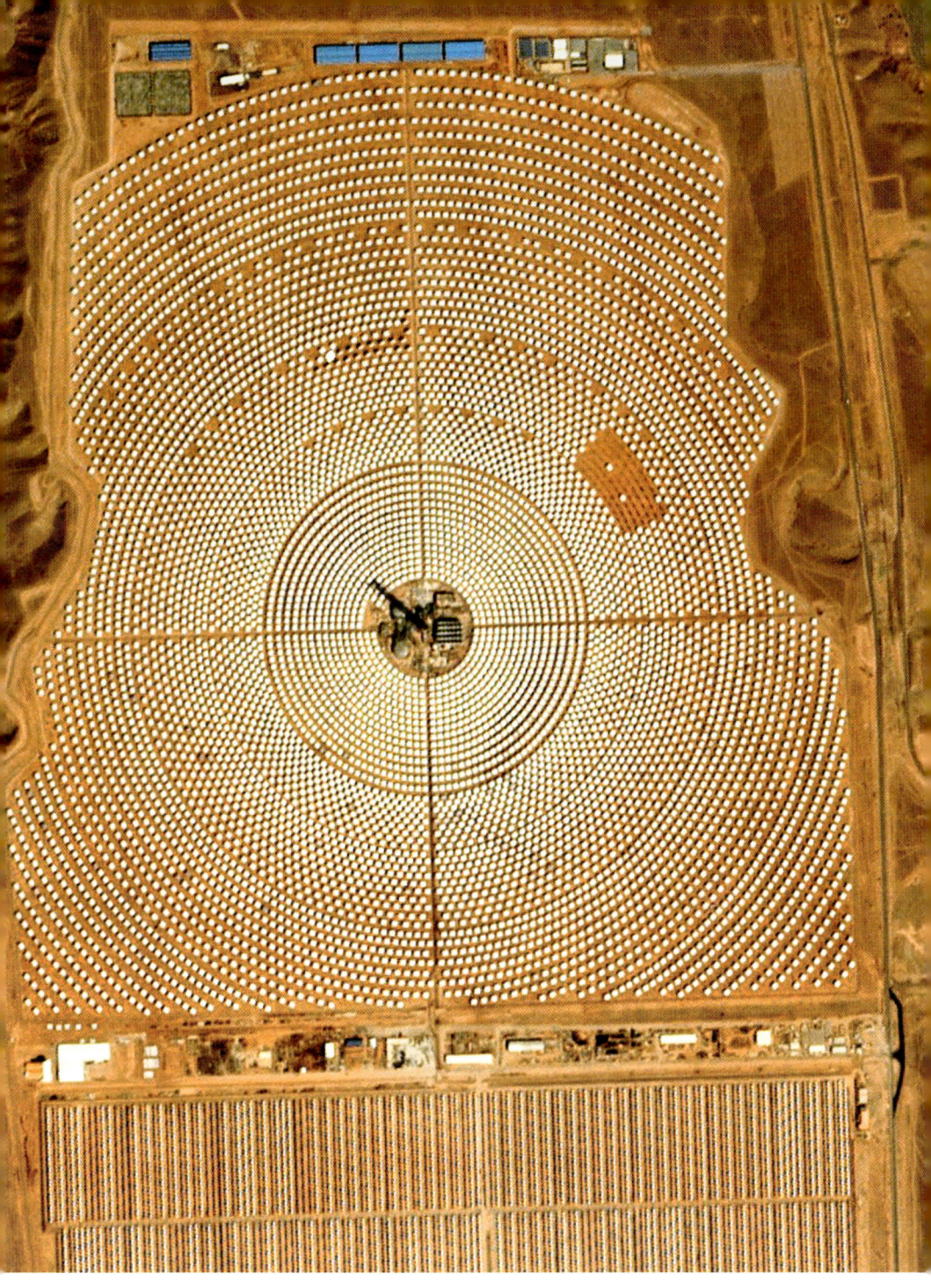

The concentric circles of a Moroccan solar plant are visible through satellite photography. According to data gathered in 2022, renewables provide about 38 percent of Morocco's electricity consumption.

1
ENERGY

An inspector stands atop a wind turbine at the Altamont Pass wind farm in California. At one point, more than 6,200 turbines were in operation here, though larger and more efficient turbines are now replacing groups of older turbines from the 1980s.

Inside the World's Most Advanced Laser Fusion Lab

M MY JOURNEY TO THE world's leading center for nuclear physics research takes me to California wine country. Few things are more iconically pleasant than cruising up and down State Highway 29, dipping into châteaus, tasting wine, and soaking in the surroundings on a sunny day, but my ultimate destination lay 60 miles (100 km) south of Napa, in a lesser known region called the Livermore Valley. It's an unassuming wine-growing region, relatively flat like Middle American farmland, but across the road from the lines of grapevines runs an eight-foot (2.4 m) fence topped with razor wire. This high-security compound, the Lawrence Livermore National Laboratory (LLNL), is guarding the holy grail of renewable energy.

I pull into an entrance gate that feels like a military base, with armed guards in camouflage uniforms checking everyone for the proper security clearance. I've gotten clearance and an escort to enter at West Gate. Thereafter, it's a short drive to the National Ignition Facility (NIF). Among the roughly 500 structures at LLNL, NIF is relatively new. Construction

Livermore uses the University of Rochester's OMEGA laser system for its fusion experiments. OMEGA's 60 laser beams can focus 40,000 joules of energy on a target in approximately one-billionth of a second.

finished in 2009, and while the facility's primary mission is to maintain the reliability, security, and safety of the United States' nuclear weapons arsenal, a major secondary objective has come out of NIF's focus on national security: the pursuit of a sustainable fusion reaction.

NIF towers 10 stories high, covers an area the size of three football fields, and houses the largest and most energetic laser in the world. Using

The Target Chamber's final optic assemblies focus the laser beams from 40-by-40-centimeter squares of light to a target as small as 0.2 millimeter in diameter.

lasers, these scientists can fuse two isotopes of hydrogen—deuterium and tritium—to form one helium atom and release a lot of energy in the process. Easier explained in a science book than done in real life. The fusion reaction itself happens in a capsule about the size of a pencil eraser, but it requires a massive structure. After all, scientists here are learning how to replicate and harness the same process that powers our sun.

Nuclear fusion energy is widely considered the endgame of renewable energy. For starters, it emits zero carbon emissions and produces significantly less radioactive waste compared to nuclear *fission* energy

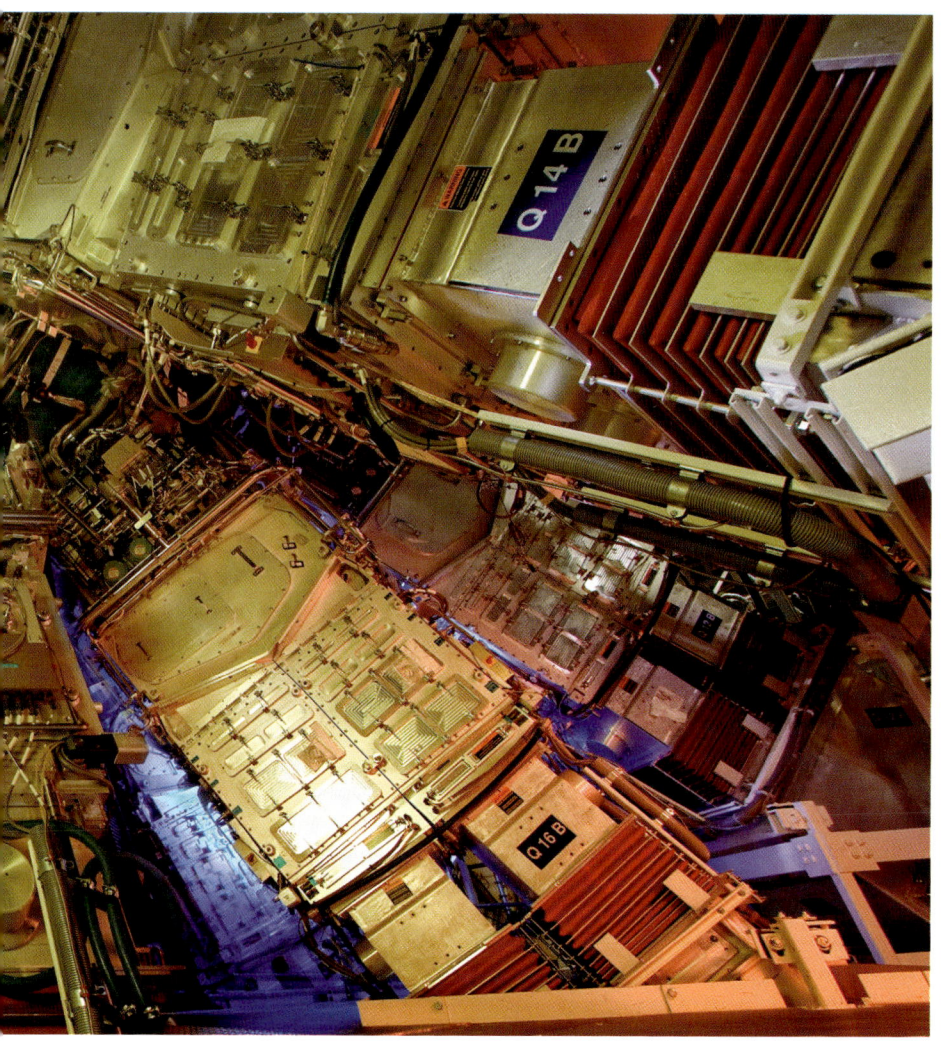

Inside one of the preamplifier support structures, which increase the laser energy pulse by a factor of 10 billion, at the National Ignition Facility (NIF)

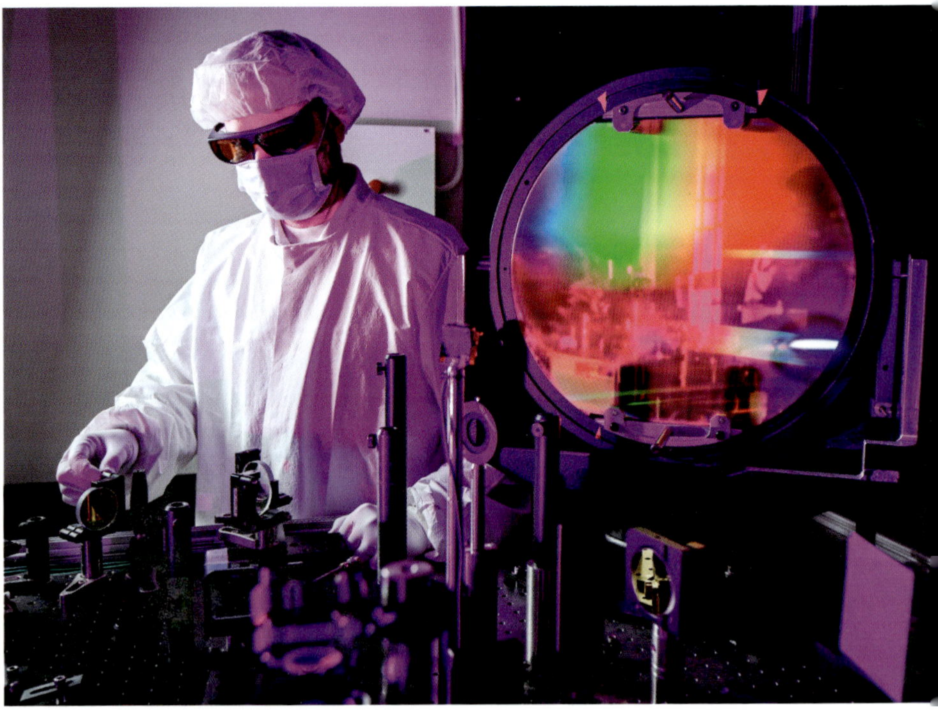

A senior NIF technician adjusts a pulse compressor, a component of the laser system at Livermore's Advanced Concepts Laboratory.

in use worldwide today. Instead, it mainly produces helium, an inert gas. Nuclear fusion packs loads more power than fission, and one of its fuel sources, deuterium, can be found in seawater. (Tritium is harder to come by, but scientists hope future fusion laboratories will breed their own self-sustaining source of the isotope.) It sounds too good to be true, but in 2022, NIF was able to produce a nuclear fusion ignition, in which the amount of energy released by the reaction exceeded the amount of energy put into it. It was a landmark achievement that Gordon Brunton, director of NIF, called a "Wright brothers moment." No kidding. Controlling nuclear fusion as a clean source of energy would be as significant as human flight.

Dr. Brunton is beaming when we meet. Born in the United Kingdom, Brunton is a gregarious, enthusiastic Scotsman who's eager to show me around NIF. On this day, NIF has scheduled maintenance. This means they're not doing any fusion experiments, or "shots," as they call them. NIF performs hundreds of shots throughout the year, and at first I'm disappointed to miss one until Brunton assures me it's not very exciting to witness in person—there is no loud boom or massive glow. It happens at the speed of light in a fraction of a second. My timing is actually advantageous: I'll get to enter restricted areas and see inside the reactor.

Brunton walks me down a dark hallway. Through windows on one side, we overlook one of the laser bays—these lasers produce the heat that will fuse the atoms together and create the reaction. In this bay, 24 square tubes sit amid dim blue lighting, each tube containing a two-by-two-foot (0.6 by 0.6 m) laser beam. It's like looking into the bay of a spaceship. Brunton describes how these lasers blast through the tubes en route to their ultimate destination, the eraser-size fuel capsule inside the spherical Target Chamber, where the reaction takes place. To access the Target Chamber, Brunton leads me through Target Bay Door 3.

The Target Chamber is an impressive structure more than 30 feet (9 m) tall, with all sorts of tubes, cables, and wires weaving in, out, and around the space. The tubes that direct the lasers line the walls, oriented toward the chamber's center. Given what I know happens here during a shot, I imagine a small figurative sun pulsating behind the capsule.

On the scaffolding above us, mechanics and engineers dressed in white clean-room suits scour the chamber for contaminants. Any speck of dust or dirt is the enemy of optics, and cleanliness and climate consistency are musts when dealing with lasers. NIF's annual budget for these tune-ups is around $300 million, and Brunton says the HVAC is their biggest monthly bill! The temperature inside NIF is always kept at 68°F (20°C), give or take a fraction, 365 days a year. Today, the mechanics are replacing some of the optic lenses that direct and focus the lasers inside the lab. These lenses are actually cut from specific crystals grown on-site in a petri dish—they take two years to grow to the size of a boulder, but the final cut lenses range in size from about six inches (15 cm) across to 20 inches (50 cm). There are more than 100,000 of them throughout the laser maze that sets up the reaction.

I also notice what looks like a giant syringe piercing its way to the chamber's middle. Brunton tells me that's Dante, the world's most comprehensive thermometer and the critical instrument measuring the heat energy produced by every shot.

Every shot starts in the control room, filled with 11 workstations modeled after NASA's Mission Control. The shot director commands the lead operator to proceed to the system shot, then the lead operator performs a roll call of the subsystems' operators and starts the countdown. This includes a NASA-like hold at T minus 5 minutes for final checks, then an ultimate 5-4-3-2-1 ... click.

The laser journey begins in the Master Oscillator Room with a pulse of low energy lasting 20 billionths of a second. This weak pulse, only about one-billionth of a joule, is split and carried on optical fibers to 48 preamplifiers that increase the pulse's energy by a factor of 10 billion to a few joules. (For more about joules and their namesake, James Prescott Joule, see sidebar on page 32.) These 48 beams are each split again into four beams for injection into the 192 main laser amplifier beamlines. These lasers shoot through the tubes we saw in the laser bay.

Guided by laser mirrors, each beam zooms through two large glass amplifiers, first the power amplifier and then the main amplifier. These amplifiers are where the pregame magic happens. They're made of neodymium-doped phosphate glass slabs. When hit with white light, the neodymium atoms are energized, then are able to amplify the low-energy laser pulses as they pass through, increasing the laser's brightness and energy.

From the main amplifiers, the beams make a final pass through the power amplifier. By now, the beams' total energy has grown from one-billionth of a joule to four million joules—an astonishing

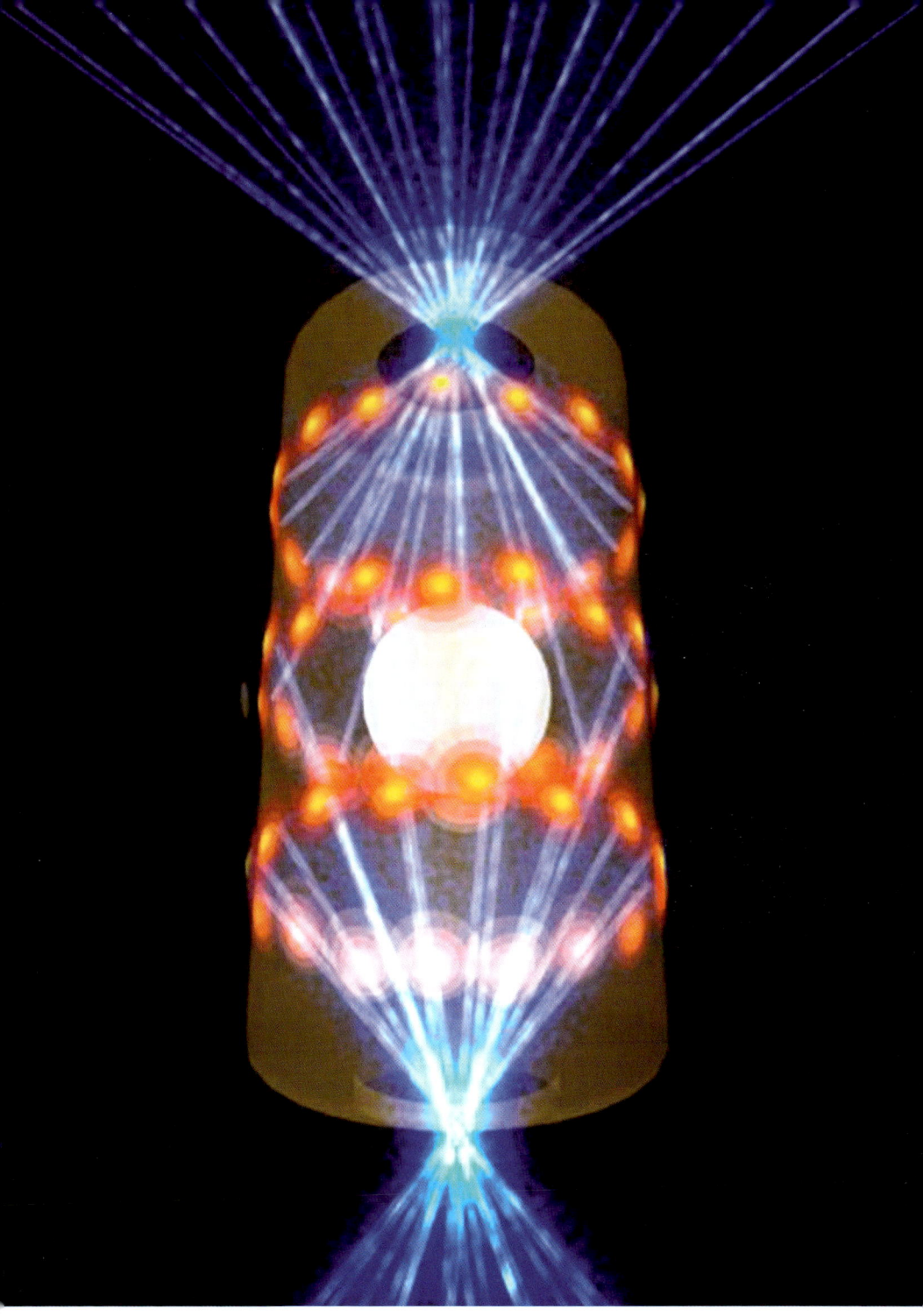

An artist's rendering of a target pellet receiving energy from laser beams at both ends of the surrounding capsule, a key process at the National Ignition Facility

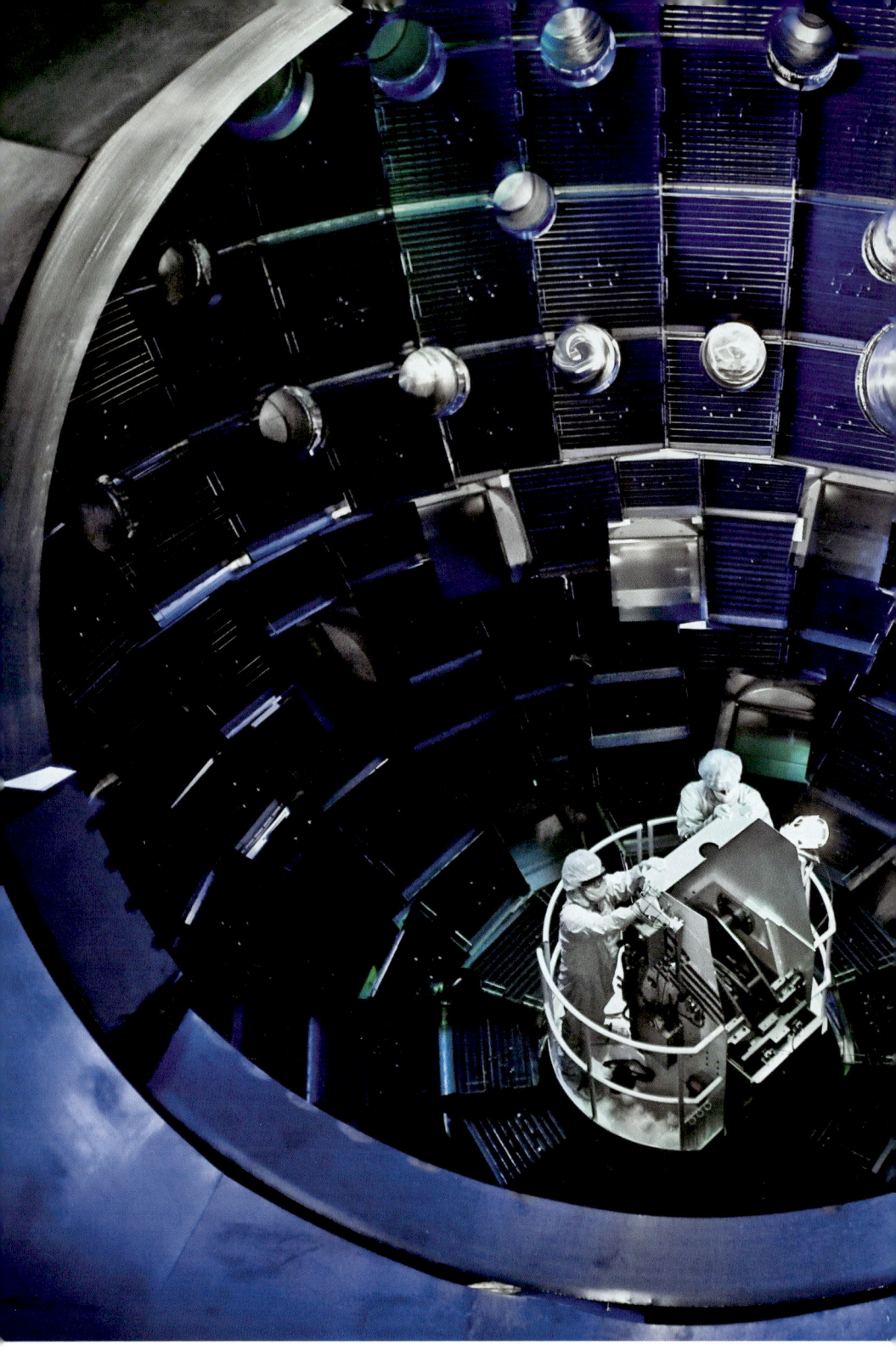

Service technicians at
Livermore ride a lift into
the Target Chamber for
inspection and maintenance.

quadrillion-fold amplification! This has happened over a distance of 1,500 meters, nearly a mile, in just a few millionths of a second.

Now at their zenith, the laser beams are redirected to enter the Target Chamber from all sides. Here, the beams shift from infrared (IR) to ultraviolet (UV) light waves—pivotal, because UV heat energy is more easily controlled and absorbed at the target location, a 10-millimeter capsule containing two forms of cryogenically frozen hydrogen isotopes suspended inside a cylindrical x-ray "oven" called a hohlraum. There, finally, the laser beams converge and unleash their energy, generating a bath of x-rays that implodes and ignites in the heart of the target sphere erupting at 5,400,000°F (3,000,000°C)—the culmination of an orchestrated symphony of precision and power.

NIF's ambitious undertaking to produce controlled ignitions of fusion reactions is at the forefront of scientific exploration for clean energy and national security. Their process is one of several that could achieve nuclear fusion energy, but all attempts at fusion have a common challenge: How do you control all of that heat without vaporizing the structure in which the fusion is occurring?

Dr. Brunton believes it's just a matter of grit, money, and time. He points back to his Wright brothers analogy: NIF is like the Wright Flyer at Kitty Hawk, a wood, wire, and cloth aircraft that would one day beget the transatlantic jetliner. One day, the Boeing or Lockheed Martin of nuclear fusion energy will turn NIF's accomplishment into a power plant—but we're a long way off from a modest fusion plant contributing energy to any grid. At least we are one step closer thanks to NIF's perseverance. As of this writing, they have achieved nuclear fusion ignition an additional seven times since 2022.

As climate change escalates to a global crisis, clean and sustainable energy has never been more urgent. Wind and solar energy are growing, but are still small pieces of our energy landscape, and that makes achieving limitless renewable fusion energy a precious goal. Waiting another 10, 20, or 100 years for unlimited clean energy is simply not an option. We need all renewable solutions put in the game now, and nuclear is the queen on that chessboard. It wouldn't be an incremental step; it would possibly be the final step to a healthy, stable planet. That's why it's at the top of our list as the next big thing.

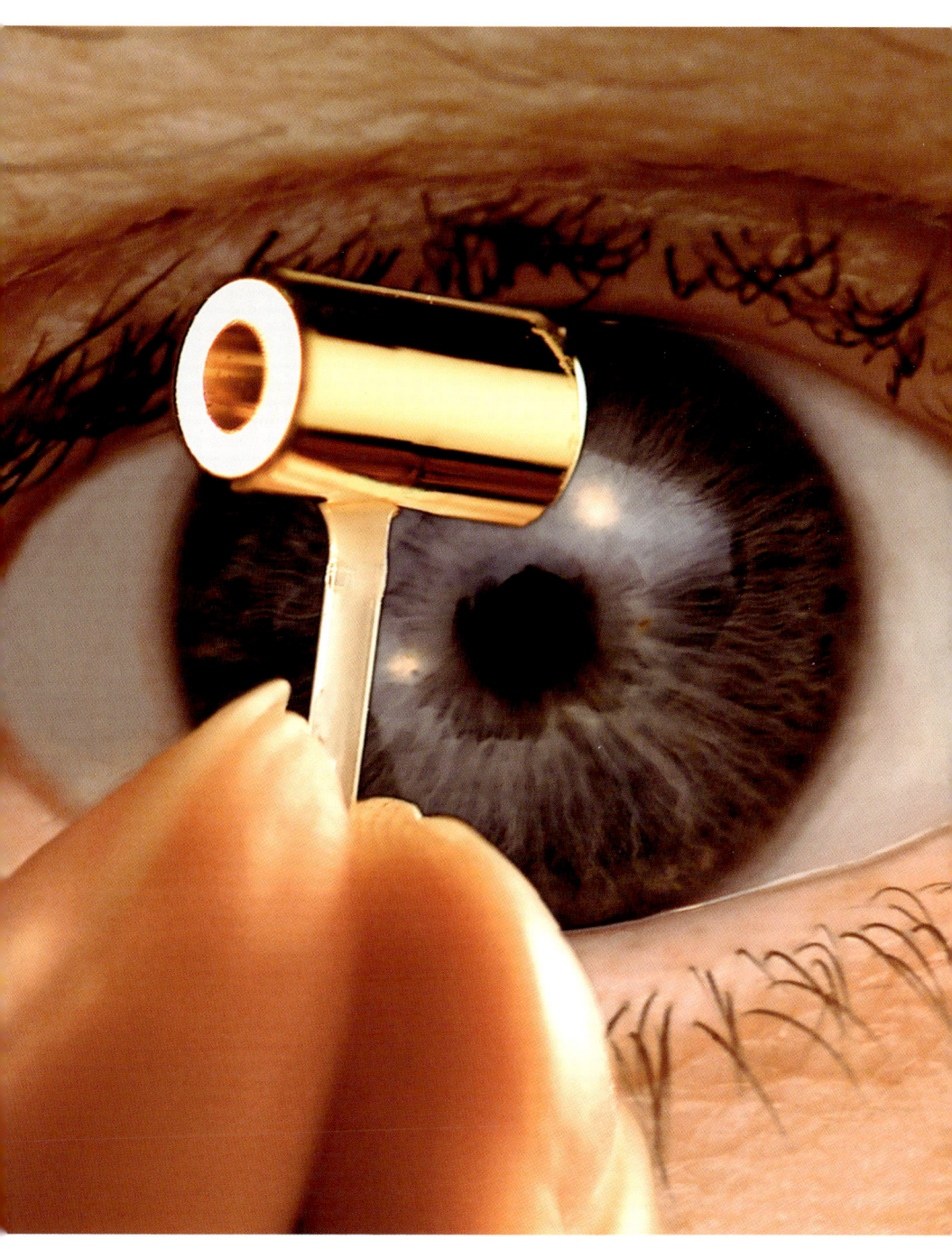

This cylinder contains a fusion capsule. The cylinder is about the size of a pencil eraser. The fuel capsule inside is the size of a pea.

1.

The **Basics** of **Energy**

W

WE OFTEN HEAR and talk about energy, but it can be a tricky thing to pin down. The fusion reactor at Livermore could one day become part of a massive overhaul for the way we run our world's energy system. Let's dive deeper into the real science of what energy is and how it works.

"Energy" is one of those terms that has a lot of different meanings in colloquial speech but a very precise definition in the sciences. We can talk about an energetic athlete or an energetic puppy, or we can compliment a speaker on an energetic speech. While those uses of the word convey a qualitative idea, you would be hard-pressed to know how much energy the athlete, puppy, or

As this train moves through Budapest, Hungary, its kinetic energy exerts a force on the track beneath it.

speaker exhibited. Energy in the sciences is talking about something very specific, yet as humans, we more often *feel* energy than become aware of it technically. So if we want to talk about energy with precision, we need a scientific definition.

ENERGY, EVERYWHERE

ENERGY IS DEFINED as the ability to do work, and work is done when a force acts over a distance. If you pick up this book, for example, you have to exert a force to lift it. If you want to move the book upward, you have to exert that force over the distance you want to lift the book. In the jargon of physics, the amount of force multiplied by the distance the book moved would be the work done or, in equivalent terms, the energy expended, when you lifted the book. In a rough way, then, you can think of energy as the stuff that is needed to make things move.

So how does this abstract statement help us deal with issues like the energy needs of an advanced technological civilization like our own? And what does that have to do with the fusion reactor at Livermore? We can start answering this question by noting the three basic facts about energy: (1) There are many forms of energy; (2) different forms of energy can be converted into one another; and (3) the total energy of an isolated system never changes. (In the jargon of physics, we say that energy is conserved.)

▶ Kinetic Energy

THINK ABOUT A RIFLE BULLET speeding through the air. Ignoring the resistance of the air, we would say that the speeding bullet is not exerting a force and, hence, is not expending energy. When the bullet hits its target, however, a force will be exerted over a distance as the bullet comes to rest. Consequently, we say that the speeding bullet has the ability to do work (or expend energy). This ability is associated with the fact that the bullet is moving, and we call this kind of energy kinetic energy—energy that is expended when the object in motion stops.

Every moving object has this sort of energy. A thrown baseball exerts a force over a (small) distance when it compresses the catcher's baseball mitt. A moving car when brought to a stop exerts a force on its brake

Sixty percent of the renewable energy of Kaprun, Austria, comes from hydroelectric power, including from the Limberg hydroelectric concrete dam.

Coal from the beds of this German lignite mine could have a carbon content up to 35 percent. It is used mostly for steam-electric power generation.

J James Prescott Joule (1818–1889), an English scientist and brewer by trade, conducted the most famous experiment demonstrating that heat is a form of energy. His apparatus was simple, as shown. A weight was raised to a height and then allowed to fall, and this caused paddles to turn, agitating the water in the container. Because of this agitation, the temperature of the water would increase. The amount of energy associated with the falling weight is easy to calculate and, provided you can measure the temperature of the water accurately (much easier said than done), you can calculate the heat that the churning paddles added to the water.

Joule's experience in brewing gave him more experience with measuring fluid temperatures than many of his scientific colleagues, and so, he was able to determine the water temperature to a fraction of a degree. As a result, he produced a more precise measurement than any of his peers could. His conclusion was that 772 foot-pounds of mechanical energy were needed to increase the temperature of the water by one degree Fahrenheit.

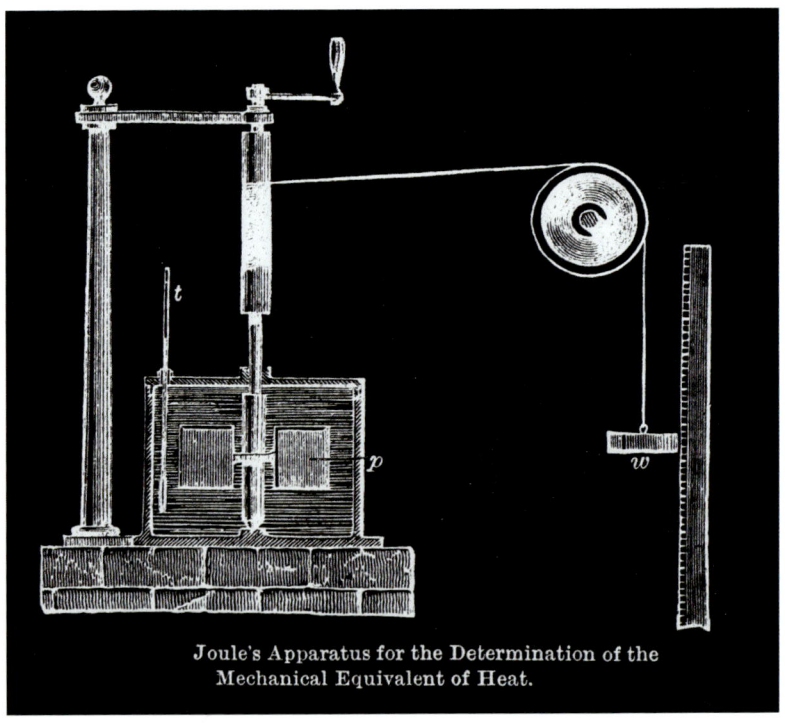

Joule's Apparatus for the Determination of the Mechanical Equivalent of Heat.

Joule's device uses a cord to rotate the paddles as the weight descends.

assembly. A galloping horse uses its muscles to exert a force on the ground when we rein it back. The point of these examples is to show that *every* moving object has kinetic energy, because every moving object exerts a force over a distance when it stops.

▶ Gravitational Potential Energy

NOW THINK OF A ROCK balanced on top of a cliff. The rock isn't doing any work, but we know that if it fell, it would exert a force over a distance when the ground stops it from falling farther. We say that the rock has potential energy—it could, under certain circumstances, exert a force over a distance and hence expend energy. Because the rock has this property due to Earth's gravity pulling it down, we say that the rock has gravitational potential energy. (To make the effect of gravity obvious, consider that if the rock and cliff were in interstellar space with no gravity, the rock wouldn't fall if you pushed it—it would just drift off.)

Probably the most useful type of gravitational potential energy is in the generation of electrical energy by hydropower. We'll discuss the details of this process in the next chapter, but for a moment, think about what happens when a dam is built. When the dam is in place, water that would normally flow downhill begins to build up behind the dam. In essence, the incoming flow of the river (kinetic energy) is used to lift the water above the level of the valley. This means that the water now has gravitational potential energy. When we want to use that energy, we let the water fall back down and spin a turbine, which, in turn, produces electricity.

▶ Chemical Potential Energy

THE ROCK AND WATER in the previous examples are capable of doing work (that is, they possess energy) because they are located in the gravitational field of Earth. Neither one would work for a system in deep space, far from any effects of gravity. Other kinds of forces, however, can give rise to potential energy. For example, in part II, we will talk about electrical charge, and particularly about the fact that electrical charges exert forces on one another, just as massive objects like Earth exert gravitational forces on objects near their surface.

All the known material objects are made from atoms, and atoms are held together by electrical forces. Basically, negatively charged electrons are attracted to a positively charged nucleus, and this holds the atom together.

Having made this statement, we need to emphasize that the common picture presented of the atom, in which electrons circle a nucleus in the same way that planets circle the sun, is not really correct. In fact, electrons are not like billiard balls or planets. According to the rules of quantum mechanics, it's much better to think of electrons as waves sloshing around orbits at specified distances from the nucleus of the atom.

Wave or particle, the distance between the electrons and nucleus of the atom determines how much energy it takes to put the electron in place or the amount of energy that would be released if the electron were allowed to fall into the nucleus. An electron in orbit, in other words, is similar to the balancing rock—both have a potential energy that depends on their position in a field, gravitational or electrical.

So what happens when two of these balanced atomic systems come near one another, as they do whenever there is a chemical reaction? The two

A masked worker protects himself from dust as he sweeps the floor of a coal-burning power plant.

(or more) atoms involved in any chemical reaction will begin with their electrons arranged in some pattern that corresponds to a total electrical potential energy, which we can calculate by adding the electrical potential of all the individual electrons in each atom. After the chemical reaction has taken place, new atoms with electrons will arrange in a new pattern. This new pattern, like the initial pattern, will have some total electrical potential energy corresponding to the new arrangement of electrons. If the final electrical potential energy is less than the initial electrical potential energy, the extra potential energy has to go somewhere. In many cases, that extra energy comes off as heat.

Now think about a gallon of gasoline sitting in the fuel tank of a parked car. Clearly, the gasoline is doing no work and hence is expending no energy. The carbon and hydrogen atoms in the gasoline have some total electrical potential energy, as will the atoms that are the products of letting the gasoline combine with oxygen in the atmosphere (a process we call burning). The heat generated by this shuffling of atoms and electrons will, in the end, enable your car to move. We say that the gasoline has chemical potential energy. The same could be said about a pile of coal or a stack of firewood, and even that frozen hydrogen at Livermore. That hydrogen has *a lot* of chemical potential energy—that's why fusion reactions are so attractive to the world's top scientists.

EXCESS ENERGY MEANS HEAT

ONE OF THE MAJOR challenges for the engineers at Livermore—and all engineers trying to achieve a nuclear fusion reaction—is heat control. Heat is another, somewhat unexpected, form of energy, and a nuclear fusion reaction generates a ton of it. Think about it: The heat from fusion reactions is precisely why we can feel the sun's rays from Earth. But for a long time, we didn't understand heat as energy. Up until the 19th century, people argued that objects contained a fixed amount of a fluid they called caloric, and that this fluid was released when an object was burned. A good fuel, like coal or wood, contained a lot of caloric while a poor fuel, like a rock or piece of ice, doesn't have much caloric.

Today, we understand that materials are made of atoms and molecules that are constantly moving. What we perceive as temperature has to do with how fast those atoms and molecules are moving—the faster they move, the higher the temperature. Looked at this way, heat is simply the kinetic energy of atoms and molecules in a substance. It has nothing to do with any fluid moving in or out. When you heat an object, you are simply increasing the kinetic energy of its atoms.

▶ The Cannon Factory

ONE OF THE FIRST people to think quantitatively about the nature of heat was a scheming American military officer named Benjamin Thompson (1753–1814). Thompson had a head-spinning career. He actually sided with the British during the Revolutionary War, serving as a spy for the British military and in the army that occupied New York City. When he realized that the British would lose the war, he abandoned his wife and children and fled to England, where he was knighted by King George III. A few years later, he had to flee England because he was accused of spying for the French (and let's face it—given his history, he probably was). He eventually wound up in the employ of the elector of Bavaria where, among his many duties, he was in charge of a factory that manufactured cannons.

Part of the manufacturing process involved mounting a cannon on an apparatus that rotated it and running a cutting tool back and forth in the barrel to make it smooth. Thompson noted that if the cutting tool was dull, the rotating cannon heated up, and the longer he kept the barrel rotating, the hotter it got. He argued that if he kept up the rotation, the barrel would continue heating—it would never run out of any stored caloric. To make his point, Thompson actually set up a boring machine in a large container of water and ran the machine until the water boiled!

Today, we would call the interactions between the atoms in the cutting tool and the atoms in the cannon barrel friction, which caused the atoms in the barrel to move faster—a phenomenon we experience as an increase in temperature. The heating had nothing to do with any stored fluid—it simply followed from the atomic interactions.

It wasn't long before scientists caught on to the idea that heat was just another form of energy. The question then turned to finding what that equivalence meant in a quantitative sense—how much coal would you have to burn to lift a weight a certain distance, for example? There

A SpaceX Dragon craft reenters Earth's atmosphere. This family of spacecraft primarily runs cargo missions and crew members to the International Space Station.

were a number of attempts to work out this sort of number, but the most famous comes from the English scientist James Prescott Joule (see sidebar on page 32). In fact, Joule was so proud of his measurement that when he died, he had his tombstone carved with the number 772.55—his 1878 calculation for the amount of foot-pounds of work needed to raise the temperature of a pound of water by one degree Fahrenheit. At the time, this was known as the mechanical equivalent of heat, or the amount of work that produces a calorie of energy. Scientists today use Joule's name to describe a unit of energy, as when the Livermore scientists report a beam increasing in energy from a billionth of a joule to four billion.

THE WORLD'S MOST FAMOUS EQUATION

BEFORE ALBERT EINSTEIN published the theory of special relativity, including $E=mc^2$, mass and energy were thought to be two separate things—unconnected reservoirs, each self-contained. It was thought that you couldn't change the amount of mass in an isolated system, nor could you change the amount of energy. What Einstein showed and expressed in that elegant equation was that the two reservoirs were, in fact, connected. You can convert mass into energy (we do this in nuclear fission reactors all the time), and energy into mass (as we do in high-energy particle accelerators, where the kinetic energy of speeding particles is used to create new forms of matter).

The equivalence of mass and energy has become a concept used in many branches of science. It is the driving force behind the generation of electrical energy in fission reactors and, as the opener to this section indicates, it will also play a role in the development of fusion reactors—a long-term (and virtually inexhaustible) source of energy in our future.

▶ The Interchangeability of Energy

IMAGINE TAKING A roller-coaster ride. Your car is pulled up to the top of the starting point. In this location you have no kinetic energy (because you aren't moving), but you have gravitational energy (because you are up high). When your car reaches the bottom, on the other hand, it has kinetic energy (because it is moving), but no gravitational potential energy (because it is at ground level). From a simple amusement park ride, you have discovered one of the most important aspects of energy: It may come in many forms, but any one of these forms can be converted into any of the others.

In principle, these conversions can be made without losing any of the energy involved. There is, however, one important exception to this rule. It turns out that if we want to convert heat into useful work, the laws of

CAN WE CREATE ENERGY FROM NOTHING?

Empty space isn't really empty. The laws that govern the subatomic world (called quantum mechanics) tell us that the vacuum is actually full of particles popping into existence and disappearing. They also tell us that once two particles are in contact with each other, they stay connected through a mechanism known as entanglement. In 2023, physicists at the University of Waterloo experimented with a theory from Masahiro Hotta, a Japanese physicist, to exploit a loophole in these laws and do something that seems impossible: Pull energy out of the vacuum.

First, the researchers started with two entangled carbon atoms. These atoms didn't have any extractable energy, yet their energy was still oscillating microscopically. When the vacuum around one carbon atom gained energy, the physicists adjusted their detectors to observe its counterpart, which also gained energy. Thus, when observing only the second atom, it appears that we are pulling energy out of the vacuum. But we are actually moving energy from one carbon atom to another—a process that defines a more futuristic process: quantum teleportation. Researchers are hoping that this work can be a key step in improving error correction for quantum computers.

A high-bandwidth magnetometer from Sydney's Q-CTRL laboratory

We convert the wind's kinetic energy into electrical energy using things like windmills, and new storage solutions for times when the wind doesn't blow are emerging.

It takes a massive force to counteract the kinetic energy of a roller coaster to stop the ride.

physics require two things: (1) To get any work from the system at all, we need to be able to take some heat energy from a hot reservoir, and (2) some of the initial heat energy has to be dumped into a cold reservoir, and once the heat is dumped, it can no longer be used. As we're about to see in part II, this requirement means that if we generate electricity by burning coal, the basic laws of physics force us to throw away some of the available energy. That's one of the consequences of using fossil fuels as a primary energy source.

▶ The Conservation of Energy

SUPPOSE YOU HAD SEVERAL different places you could put your money—a checking account, a savings account, a money market account, and so on. Suppose further that you had $100. You could distribute the $100 dollars in those accounts in all sorts of ways—you could put it all in one account, the same amount in every account, more in some accounts than in others, and so on. No matter which scheme you chose, however, you would always have exactly $100. Using the language of physics, we would say that the total amount of money is "conserved."

Energy is like that. As we have seen, there are many different types of energy (just as there are many different types of bank accounts). Energy can be shifted back and forth between the different types, just as money can be shifted from one account to another. And just as shifting your money around doesn't change the amount you have, the total amount of energy in an isolated system stays the same. In this universe, there is no free lunch. Everything costs something.

Let's look at the roller-coaster ride we just discussed to see how this works. At the top of the ride, all your energy is in the gravitational potential energy account. As you start down the slope, you begin transferring energy from that account into the accounts labeled kinetic energy and heat (because the friction of the wheels on the track generates some heat). By the time you reach the bottom, the gravitational potential energy account is empty, and all of its funds are split between the kinetic energy account and the heat account. These rules are crucial to building the energy technologies of the future: In a world where all energy is conserved, we have to ensure we are putting energy into the right accounts—and sometimes, that's very difficult to do.

CHAPTER

2

The **World Energy** Picture

W **WE USE ENERGY** for so many things in our lives. It heats our homes, moves us around, runs our factories, and grows our food. The first law of thermodynamics, the law of conservation of energy, states that energy can neither be created nor destroyed. From that, we can make a definitive statement about all the energy we use: It has to come from somewhere. As we have pointed out, there is no free energy lunch anywhere in the universe—even in a vacuum! We have to pay for every bit of energy we use.

Most of our energy comes from the sun. Sometimes we use sunlight directly to generate heat or electricity, and sometimes we use sunlight stored

This open-pit lignite mine in Germany shows how the mining of fossil fuels can scar the planet.

in fossil fuels like coal and petroleum. Indirectly, we use the sunlight that creates the global temperature differences that make the wind blow too. The sun's energy stores are vast, but the world is facing a great problem: How can we tap in to all these forms of solar energy without damaging our ability to manage the excess heat we create? Solving this dilemma is key to a sustainable energy future.

THE SUN'S ENERGY

DEEP IN THE CORE of the sun, nuclear reactions take place whose net effect is to convert four hydrogen nuclei into a helium nucleus. The four hydrogen nuclei weigh a little more than the helium nucleus, and the difference in mass is converted into energy in the form of intense x-rays and gamma rays (remember $E=mc^2$). This energy bounces around inside the sun as it slowly makes its way to the surface—a process that takes tens of thousands of years or more. In fact, the sun has many layers in its structure, and the movement toward the surface is a little different in each of them—this is one reason it takes so long to get there. By the time the energy reaches the surface, it is no longer in the form of x-rays, but has cooled down to visible light. Once free of the sun's mass, the light travels through the vacuum of space between the sun and Earth in about eight minutes.

The next time you feel the warmth of the sun on your face, then, you may want to think about the fact that Neanderthals still walked on Earth when that heat was produced in the sun's core! This magnificent timeline is why we can think of the sun as a renewable energy source: It's going to provide heat and light for billions of years to come.

▶ Renewable Versus Nonrenewable Energy

EVEN IF YOU USE THE ENERGY in sunlight that falls on a patch of ground today, more sunlight will land on that patch of ground again. Similarly, wind blowing through a mountain pass to run a windmill today will blow again in the future. These are both examples of renewable energy sources.

On the other hand, coal burned to produce electricity or gasoline to run a truck is not replenished on any meaningful timescale. These are nonrenewable energy sources—they took up to hundreds of millions of

These solar photovoltaic panels in Vietnam are collecting energy first produced by the sun tens of thousands of years ago.

years to form and will take millions of years to be replenished. Though fossil fuels are the subject of debate amid the world's changing energy picture, they are still central to global energy consumption. They're simply the most convenient form of energy given our current infrastructure.

▶ Nonrenewable Fossil Fuels

FOSSIL FUELS REPRESENT sunlight captured millions of years ago and transformed by slow geological processes into the coal, oil, or natural gas that we use. Eons ago, sunlight arrived at our planet. Plants converted it into vegetable matter that grew, died, decayed, and dropped deep underground. Algae and other organic materials floating on the ocean surface absorbed sunlight and settled down as well. All these processes ultimately led to the product we call petroleum.

Today, this long-ago captured sunlight is the main energy source driving our societies.

The problem, of course, is that this form of stored sunlight is neither renewable nor green. No process will replace that lump of coal on a human timescale. What's more, that burning releases carbon dioxide into the atmosphere. We need ways to tap sunlight as soon as it arrives on our planet, and scientists and researchers are finding ingenious ways to do so. The critical question is whether humanity at large can adopt these techniques at a scale that makes our energy picture cleaner and more efficient.

▶ Green Versus Non-Green Energy

THE ADDITION OF carbon dioxide to Earth's atmosphere is seen as a major problem in our energy future. By definition, a so-called green energy source is one that does not add to the atmosphere's carbon dioxide load. Wind, solar, and nuclear energy all fall into this category. Coal, oil, and natural gas do not. Other examples of green energy sources include most biofuels, which we'll discuss soon. In these systems, sunlight is converted into chemical fuels through photosynthesis in plants. The carbon that enters the atmosphere when fuel burns is balanced by the intake of carbon as plants grow, making biofuel "carbon neutral."

Nuclear power poses a somewhat difficult classification problem. Generating electricity from nuclear reactions produces no carbon dioxide, and hence, nuclear must be considered a green energy source. On the other hand, current fission reactors require uranium as a fuel source, and uranium, like coal, is a naturally occurring mineral that must be mined and, once used, cannot be used again. Thus, nuclear fission energy has to be classified as a nonrenewable but green form of energy. (Keep in mind the distinction between nuclear fission energy and nuclear fusion energy like that at Livermore. Once fusion reactors can breed their own tritium, they will be both renewable and green.)

TURNING HOOVER DAM INTO A BATTERY

Hoover Dam's 17 generators already bring power to 1.3 million homes, but as the United States debates a greater shift to renewable energy, engineers want to turn it into a battery that can provide electricity when the sun isn't shining and the wind isn't blowing.

Here's how it would work: An underground pipeline would connect Hoover Dam's main reservoir, Lake Mead, to a proposed pump station 20 miles (32 km) downstream. During the day, water would pump upstream into Lake Mead while wind and solar support the grid. At night, water would fall back into the pump station to deliver an electricity boost during peak hours when wind and solar are less active. This would theoretically be more reliable than the lithium-ion batteries many renewable stations use to store excess energy today—those last just five to 10 years. Meanwhile, Hoover Dam's been pumping since 1936. Though scaling hydroelectric power might be difficult, this solution could be a model for integrating our current systems into the renewable picture.

Hoover Dam provides power for utilities in Arizona, California, and Nevada.

The most common biofuel today is wood, though researchers are investigating other sources, like algae cultures.

OUR CURRENT ENERGY USAGE

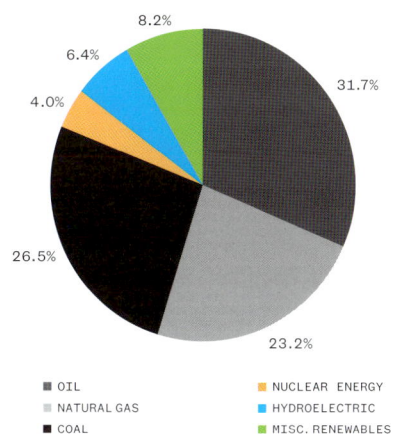

8.2%

6.4%

4.0%

31.7%

26.5%

23.2%

- OIL
- NATURAL GAS
- COAL
- NUCLEAR ENERGY
- HYDROELECTRIC
- MISC. RENEWABLES

IN A SCHEMATIC OF where the world got its energy in 2023, there are several important points to notice. The most striking is the fact that more than 80 percent of the energy used worldwide comes from fossil fuels—coal, oil, and natural gas. These fuels play different, but essential, roles in modern societies. Coal is primarily used to generate electricity. Oil is used to power modern transportation systems, while natural gas is used primarily for residential heating and electricity generation. What all these fuels have in common is that extracting useful energy from them creates methane and carbon dioxide, the prime instigators of climate change. Thus, as we seek energy sources that do not add to climate change, we must first question how to change the reliance of modern societies on fossil fuels.

This is a difficult problem, not only because so much of our world already depends on them, but also because fossil fuels are abundant, relatively easy to recover, and high in energy content. In addition, recent technologies like fracking have made the discovery and exploitation of new sources of fossil fuels relatively easy and cheap. On the other hand, we do have other sources of energy available. So how are different countries managing this transition? Most developed countries have adopted a goal of being carbon free or carbon neutral by 2050. Whether this is a reasonable goal remains a subject of debate, but it reflects the growing political consensus about the necessity of reducing the amount of carbon dioxide human activities introduce into Earth's atmosphere.

▶ The Tale of Three Countries

WE CAN FURTHER examine the energy landscape by zooming in on three countries: the United States and China, the two largest emitters of carbon

A coal storage site in northern Germany; about 18 percent of the country's energy came from coal in 2023.

Gas and oil storage tanks in the Netherlands illustrate how, together, gas and oil made up 77 percent of the country's energy consumption in 2023.

dioxide in the world, and Denmark, generally considered to be one of the world leaders in transitioning to a green economy. The differences between these countries are striking. Denmark already generates an incredible 80 percent of its energy from renewable sources—with wind, hydropower, and biofuels leading the way. China, on the other hand, generates more than half of its energy from coal, which is generally considered to be the least green of the fossil fuels. This is a strange fact, because China is also a leading supplier of solar energy equipment to the rest of the world. The United States, on the other hand, has shifted much of its electricity generation away from coal to natural gas, which emits about half as much carbon dioxide as coal when used in this way. Today, the United States emits less carbon dioxide than it did before this transition away from coal.

But even Denmark—the poster child for our green future—still generates more than 10 percent of its energy from fossil fuels. So even if the rest of the world caught up to Denmark, and even if everyone drove electric vehicles (EVs), plenty of those cars would still charge using electricity from fossil fuels. Not quite the green future that people dream about.

The other important lesson from these case studies is that the type of renewable energy a country adopts depends critically on its geography. Denmark is a small country with a lot of coastline, and so it has ample space for offshore wind farms and doesn't need to use agricultural land for energy generation. On the other hand, Denmark's northerly location means it has very little solar energy capacity.

The United States and China, meanwhile, both have long coastlines and areas of undeveloped land suitable for large arrays generating solar energy. This geographical reality is reflected in the fact that both have roughly equal percentages of wind and solar energy in their energy generation portfolios. So how do we go from a world that largely runs on fossil fuels to one that runs on renewable energy sources? The diversity in approaches between the United States, China, and Denmark shows that despite variations in political systems and geography, there are multiple paths to renewables.

▶ Converting to Hydro Energy

ONE COMMONALITY ACROSS the United States, China, and Denmark is that all three countries generate only a small proportion of their energy from hydroelectric systems. In essence, these systems depend on the sun's energy, which drives Earth's water cycle, evaporating water from the world's oceans and lakes and pulling it into the atmosphere. The water then falls as rain at higher elevations. Its journey back down to the sea is interrupted by dams, so that large amounts of water are stored at high elevations—developing gravitational potential energy. The stored water, allowed to fall, can spin a turbine to generate electricity.

SHOULD WE TURN WOOD INTO ELECTRICITY?

Green energy scored an apparent win in 2009, when the European Union (EU) classified wood pellets as a renewable biofuel. A lot of that wood comes from forests in the southeastern United States, where fast-growing evergreen trees are regularly harvested like any other crop. In 2023, the EU consumed 24.8 million metric tons (27.3 million U.S. tons) of this biofuel.

But in recent years, scientists and environmentalists have questioned if wood pellets are really as renewable as they're claimed to be. According to the Natural Resources Defense Council, wood pellets produce more CO_2 per unit of electricity produced than coal. In 2018, about 800 scientists signed a letter calling for the end of wood-pellet burning, and in 2022, the EU began limiting wood pellet subsidies but declined to declassify them as renewable. The hope is that as solar and wind grow more affordable, this controversial biofuel will go by the wayside.

Steam erupts from a German power station where wood becomes biofuel.

Hoover Dam in the United States, the Aswan High Dam in Egypt, and the Three Gorges Dam in China are all examples of hydroelectric facilities.

Generating electricity in this way has many advantages. For one thing, it is completely green. (Well, that is to say, it's green once the dams are built—the massive amount of concrete used in dams produces a lot of carbon dioxide.) For another, hydropower can be turned on and off quickly, and can therefore match demands placed on the electric grid. The two problems with hydroelectric power are that it requires specific terrain—nearly everywhere on Earth where a dam could be built already has one in place—and that dams negatively impact the surrounding environment. Hydroelectric power will no doubt play a role in our green future, but it really isn't capable of being expanded to meet the needs of our growing population.

▶ Converting to Biofuels

CHINA, THE UNITED STATES, and Denmark each have a category of energy generation connected to burning organic material, typically materials derived from plants: biofuels. Perhaps the most familiar of these energy sources is ethanol, used to drive automobiles and trucks. In the United States, for example, the sugars in corn (maize) are converted into ethanol, which is then mixed with gasoline. In fact, almost all gasoline sold in the United States is blended with 10 percent ethanol. In Brazil, ethanol is derived from sugarcane and can be used as fuel directly. Some vehicle fuels are even derived from peanuts.

Because the energy of ethanol comes from combining oxygen from the air with carbon in the molecule to produce carbon dioxide, you may be wondering why ethanol is considered a green fuel. Although burning ethanol introduces carbon dioxide into the atmosphere, the carbon in plants was pulled out of the air while the plant was growing. Thus, growing and burning the plants, in the end, introduces no new carbon into the atmosphere—what goes into the air by burning was pulled out of the air during the plant's growth. In the end, the cycle is carbon neutral. This argument plays an important role in the rise of wood as an energy source (see sidebar opposite).

Wood supplies about one percent of the energy used in the United States. This might seem strange, as we tend to think of wood as an old-fashioned energy source, powering things like fireplaces and antique steam loco-motives. But today, wood has a few modern uses as an energy source.

One is in paper production. Once wood fibers have been incorporated into paper, the paper has to be dried—a process that consumes a lot of energy. Paper companies long ago learned to use wood scraps—the detri-tus of the paper production process itself—to supply the heat for drying. This tried-and-true industrial use of the material largely accounts for the surprising one percent in the list of energy sources.

▶ Fuel or Food?

ONE ASPECT OF THE production of biofuels is potentially problematic, how-ever, and this is the competition between the use of crops to produce fuel versus to feed people. A given plot of land can produce only so much corn. If that corn is used to run cars, it can't be used to feed people or animals. Because the human population is expected to reach about 10 billion by the end of this century, the conflict is serious.

But researchers are developing a new technology to deal with this problem. The basic idea is to find a way to convert the inedible parts of a plant to fuel while reserving the edible parts for food, but that's easier said than done. Though converting the sugars found in corn or sugarcane to ethanol is relatively easy, converting the cellulose fibers that make up the rest of the plant is much tougher. Only a few organisms can digest these fibers—termites are a well-known example—but if we could do this ourselves through an innovative process, we would have a rich new etha-nol source. Finding a commercially feasible way of producing cellulosic ethanol, as it's called, remains a serious challenge to solve.

▶ Converting to Geothermal Energy

THOSE OF US WHO LIVE on the surface of the planet rarely think about one important nonsolar energy source, and that is the planet itself. The interior of Earth is very hot: Its core has a temperature between 7200°F and 11,000°F (4000–6000°C)—about the same as the sur-face of the sun. Some of this heat is left over from Earth's formation, when massive meteor impacts melted the planet all the way through, and some comes from decay of long-lived radioactive isotopes. When hard-rock miners open tunnels more than a mile (1.6 km) deep, they run air conditioners to cool the tunnels before workers can enter.

If we can access hot rocks relatively near the surface, we can drill to them, pump in water to be heated, then use the superheated water or steam to spin a turbine and produce electricity. Iceland, which sits atop one of the most geologically active regions in the world, produces more than 25 percent of its electricity using this technique. It uses the heated water to warm homes and businesses—and in Reykjavík, once the water has been used to heat buildings, it is run through underground pipes to melt snow and ice in the streets. This marriage between geography and infrastructure is the ideal for changing our current energy picture. In part III, we'll see why Scandinavia features it in so many timber sky-scrapers. Conversely, in the next chapter, we'll see why sending energy harvested from wind and solar farms to American cities proves a challenge.

These cornfields near Cairo, Egypt, could provide a vital source of carbon-neutral energy— or they could feed hundreds of people. The debate continues.

3

Can **Wind** and **Solar** **Power** the Planet?

G **GIVEN THAT WE** want to build future societies on a carbon-neutral energy system, we have to assess current technologies for being renewable *and* green. In the last chapter, we found that energy sources like biofuels, wood pellets, and hydroelectric power have potential, but could be limited by current infrastructure or the rigidity of societal norms. They might contribute to the future, but for now, solar and wind energy are generally considered to be the main contenders to play the leading energy role going forward. Both solar and wind are undergoing rapid development around the world and, with suitable caveats about the energy

Wind turbines and solar arrays in Mojave, California, spread across an area of approximately 1,765 acres (700 ha).

storage problem, are already competitive with fossil fuels in the electricity market. Turns out, each presents us with two viable but different forms right now, which means that in this chapter, we have four possible energy options to examine.

| SOLAR ENERGY |

YOU ONLY HAVE TO GO outside on a hot day to know that Earth receives energy continuously from the sun. To be precise, the energy in the incoming sunlight at the top of Earth's atmosphere averages roughly 340 watts for every square meter (about 11 square feet) of Earth's surface, but almost 80 percent of that never reaches the ground. Of this energy, about 77 watts per square meter are reflected or absorbed by the atmosphere and never reach the surface at all. An additional 23 watts per square meter are reflected, and another 163 watts per square meter are absorbed at Earth's surface. This means that we have access to about 77 watts per square meter—only enough to run seven or eight light bulbs.

This energy can be deployed in basically two ways. One is to use the solar energy directly—for example, for space or water heating in individual homes designed to capture the sunlight. Twentieth-century solar panels were designed for this household purpose as well. By far the most important future uses of solar energy, however, involve using sunlight to produce electricity, so we will focus on these innovative solar systems capable of producing electricity at a large scale.

▶ Solar Thermal Energy

THE MOST OBVIOUS WAY to harness the sun's energy is to concentrate the incoming radiation, like Boy and Girl Scouts might light a fire by using a lens to concentrate incoming sunlight on a pile of tinder. One technique used to produce large amounts of electricity involves what is essentially scaling up this old trick. This technique goes by the name of solar thermal generation.

Large-scale solar thermal installations typically have hundreds of steerable mirrors directing sunlight to a chamber at the top of a high tower. The net effect from the concentrated solar energy can be used

Some solar cell technology, such as the smallest of doped semiconductors, is light enough to rest on a soap bubble.

to boil water and produce steam to spin a turbine and produce electricity. In solar thermal systems, the heat generated by sunlight plays the same role in generating electricity that nuclear reactors and coal furnaces do.

This type of installation has clear advantages—it is a single, localized source of electricity, which makes it easy to integrate into the existing grid—but it has two major disadvantages shared by many renewable energy sources. One is that it can operate only when the sun is shining. Given that peak energy use in northern cities usually occurs at or after sunset, this is a problem. The other disadvantage is that solar energy is

---

▶ HOW TO DOPE A SUPERCONDUCTOR ◀

The most common semiconducting material is silicon, which has four electrons in its outer orbit. In solid silicon, electrons in each silicon atom hook up with neighboring atoms to form chemical bonds. These bonds can be manipulated through the process of doping.

Here's how doping works: When silicon is melted to be made into chips, another substance is added to the mix. This new substance can have more or fewer than four electrons in its outer orbit. For example, phosphorous has five electrons in its outer orbit—four of which will be used to create chemical bonds to the silicon atoms and one of which will wander off, leaving behind a positively charged atom locked into the silicon matrix and making a semiconductor doped with a positive charge. Had we doped with something like boron, which has three outer electrons, a loose electron would eventually supply the missing bond and we would have a negative charge locked into the matrix.

n-doped silicon   p-doped silicon

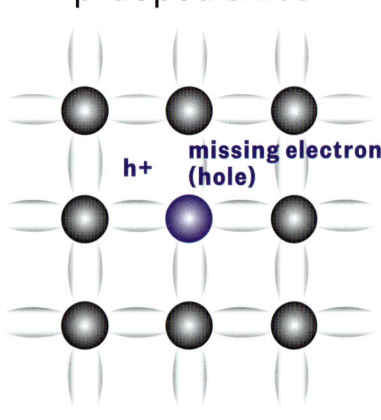

e-   extra electron   h+   missing electron (hole)

diffuse, so solar thermal installations require a lot of land, which means they are generally built in deserts far from major population centers.

One of the world's largest solar thermal power installations is the Noor Ouarzazate Solar Complex in southern Morocco, at the edge of the Sahara desert. The name Noor comes from the Arabic word for "light." The Noor power station covers 12 square miles (31 km²) of land with mirrors, and it generates about 580 megawatts (MW) of electricity. Noor's designers came up with an ingenious solution that allows it to provide electricity at night. During the day, it uses solar energy to heat a bath of chemical salts to 932°F (500°C), and at night, it uses this heat to produce steam to run turbines and keep the current flowing. It's impressive engineering, but scaling something like the Noor will be costly in terms of land. Its 580 MW output is about half of what a standard fossil fuel plant or nuclear reactor would produce.

▶ Solar Photovoltaic Energy

A MORE COMMON TYPE of solar collector uses the properties of semiconductors to convert incoming sunlight into direct electrical current. Semiconductors (discussed both here and in part II) are materials in which electrons are bound so loosely to their atoms that the electrons can be shaken loose.

The texture of a solar cell's surface can reduce reflection.

The flat black solar collectors you see everywhere—on rooftops and at remote construction sites—all use the properties of semiconductors to produce electrical current. The term "photovoltaic" means that these collectors produce electric current directly from light, without a turbine or other intermediate steps.

The properties of silicon can be manipulated to produce materials called doped semiconductors—essentially, silicon that has positive or negative charges built into its atomic structure (see sidebar on page 60). Solar photovoltaic panels are built with a thin layer of silicon doped with a negative charge laid over a layer of silicon doped with a positive charge. This arrangement creates a voltage between the layers that will accelerate electrons toward positive charges.

The idea is that when sunlight falls on a solar panel, some electrons will be shaken loose from the silicon atoms. These electrons will be attracted by the positive ions (atoms with an electric charge) and pushed into the external wire. These electrons constitute the direct current (DC) power generated.

▶ Limits on Solar Panels

AS LONG AS THE SUN is shining, it seems this simple array of semiconductors will supply us with an inexhaustible source of energy. But not quite. For one thing, not all photons from the sun have enough energy to shake electrons loose—their energy will just go to heating up the collector. This and other more technical problems limit the efficiency of solar collectors. Typically, an average solar collector converts only about 21 percent of incoming solar energy into electricity. This contributes to one of the drawbacks to using solar power—it requires a huge amount of land, covered with collectors, to make a dent in any nation's energy supply.

The advantage of the solar photovoltaic system is that once you've expended the energy needed to create the doped semiconductors, the current will flow as long as the sun is shining. There is nothing analogous to the presence of a boiler and generator in the solar thermal system. On the other hand, the solar photovoltaic system has no way of storing energy without an independent storage solution, as is done at the Noor installation, which uses a molten-salt system to store heat and produce electricity at night.

Flexible solar photovoltaic cells in Spain offer architectural and aesthetic advantages, and they may be more efficient as well.

▶ Solar Photovoltaic's Fast Rise

PERHAPS THE MOST STRIKING thing about solar photovoltaic panels is the fact that they can be used individually or in small groups. Unlike solar thermal systems, which can be used only in massive arrays, solar photovoltaic systems can be used to power isolated workstations or individual homes as well as in large arrays. You often see them, for example, powering traffic signals in remote areas where connections to the grid might be difficult and expensive. This flexibility is undoubtedly a main factor in explaining why this sort of solar energy is being adopted so quickly. In fact, because of both this and decreasing costs of solar power implementation, solar energy—particularly photovoltaic energy—has become one of the fastest-growing energy sources in the world. In 2021, almost 4 percent of the electricity generated in the United States came from this source. This is not surprising, because as large as it is, the United States has a huge potential supply of solar energy. It has been estimated, for example, that all of the country's electricity needs could be met by exploiting the sunlight falling on an area about the size of Lake Michigan.

WIND ENERGY

THE SECOND GREAT source of green energy that is expected to take us into the next century is wind. Humans have exploited wind for a long time, both for powering sailing ships and for running mills to grind grain and spices. Today, wind power is by far the greatest source of renewable energy in the United States, with more than 75,000 turbines contributing over 10 percent of the nation's electricity.

Energy from the sun drives wind power. The main force that creates wind is the difference between the amount of the sun's energy striking the Equator and the amount of energy striking the poles. If Earth didn't rotate, warm air would rise at the Equator, travel north, and descend at the poles. The surface winds in the Northern Hemisphere would then blow from north to south as air returned to the Equator.

But because Earth rotates, these wind patterns change. In fact, the Northern Hemisphere has three zones of prevailing winds—trade winds blowing east to west near the Equator, prevailing westerlies blowing west to east at the midlatitudes, and winds blowing west near the pole. The Southern Hemisphere has a similar pattern.

Thus, when we talk about the energy in wind, we can say that it comes (in a complicated way) from both the sun and Earth's rotation. Winds blow from regions of high pressure to regions of low pressure, and storm patterns, local topology, and many other factors influence them as well. Working out how the wind will blow at the location of a particular wind farm, then, can be a very complex task.

▶ Wind Turbines

MODERN WIND TURBINES bear little resemblance to the picturesque windmills of yore. In modern systems, complex machinery is needed to convert the slow motion of blades into electrical current, but this is only a small part of the physics and engineering that go into harnessing wind energy.

In the machinery needed to convert the movement of a windmill blade into usable electric current, a shaft connects the blades to a set of gears that convert the blades' slow rotation into a controlled high-speed rotation of another shaft that in turn rotates a collection of copper wires held between the poles of a permanent magnet (a magnet that retains its magnetic field without an inducing current). This rotation produces an electrical current in the wires.

To understand how a modern windmill works, let's start with some basic facts about the motion of a substance like air. When air blows over the ground, the ground tends to slow down the air's motion—it may even stop the motion completely. As you move higher in elevation, however, the interaction of air with ground—called the boundary layer in this context—becomes less important, and airflow becomes stronger and more stable. One of the first things to ensure in designing a wind system, then, is that the blades are above the boundary layer—and that is why commercial wind turbines are so tall.

The second point to consider is that the bigger the area of wind intercepted by windmill blades, the more energy the blades will be able to produce. This fact has led to a steady growth in the size of windmills. In fact, many modern machines may have blades almost as long as football fields. In 2023, China's Three Gorges Energy connected the gargantuan MySE 16-260 wind turbine to the grid. Each of the MySE's blades measures 404 feet (123 m) in length and weighs 54 U.S. tons (49 metric tons)!

▶ How Much Energy Is in a Windmill?

REGARDLESS OF ITS height and the length of its blades, a single wind turbine in operation today is unlikely to produce more than about 5 MW of energy—roughly enough to power 5,000 homes. (That's an average—the MySE is a 16-MW windmill, hence it produces more power.) This may seem like a lot, but against the energy requirements of a modern city or nation, it's a drop in the bucket. Any serious discussion of wind energy has to involve the concept of wind farms: collections of dozens and even hundreds of windmills at a single location. This brings us back to the question of airflow.

▶ Planning a Wind Farm

THE OPTIMAL CONDITIONS for extracting energy from the wind require that the flow of air past the blades be smooth and nonturbulent. Once this smooth flow has interacted with a windmill, however, there will be disturbances in the downstream flow. Think of this as being analogous to the swirling of autumn leaves on a street as a car goes by. If you are going to have a lot of windmills in a particular location, you need to place the downstream windmills far enough away from the initial machine for these disturbances to die down. Engineers adopt a rough rule of thumb that windmills need to be separated by about six times the length of their blades. Given this requirement, and especially considering the massive machines appearing in China and elsewhere, it's not hard to see that wind energy, like solar energy, requires a great deal of space.

One way to deal with the spacing problem is to make individual windmills bigger, so that we can get the same output from a smaller number of windmills. Bigger installations have other advantages, and the size of the machines has been growing steadily. There are even plans on the drawing boards for windmills that will be taller than the Empire State Building or the Eiffel Tower (although where you would put such monsters is an open question).

▶ Offshore Wind Farms

AND THIS, OF COURSE, brings us to the question of where we are going to build wind farms. Basically, we have two choices—we can build them on land or we can build them in shallow coastal waters (onshore and offshore installations, respectively.)

One of the largest offshore wind farms in operation today is Hornsea 2—an installation in the North Sea off the eastern coast of England. It consists of 165 turbines spread out over an area of 178 square miles (461 km²), generating in total about 1.3 gigawatts (GW) of electrical power—a little more than a typical coal-fired plant or nuclear reactor. The turbine blades are about 90 yards (80 m) in length, and the windmills stand more than 600 feet (180 m) above sea level at the top of their sweep.

The world's largest onshore wind farm is located in China: the Jiuquan Wind Power Base, at the edge of the Gobi desert. There, more than 7,000 turbines spread over 18,000 square miles (47,000 km²). At full capacity, it is slated to produce 20 GW of electrical power. The most recent data says it currently produces about 10 GW.

ARE WIND AND SOLAR THE FUTURE?

DESPITE THEIR PRESENT shortcomings, these four energy sources—direct solar, solar photovoltaic, onshore wind, and offshore wind—seem to be our best bets at the moment for getting to a renewable and green future. There

WILL IT FLOAT? WIND FARM EDITION

The continental shelf drops off quickly on the Pacific coast of the United States. For engineers used to building offshore wind farms in shallow coastal waters, that's a problem. How can you build a wind farm if you can't reach the ocean floor?

Engineers at GE Global Research posed the obvious solution: Make the windmills float. Their proposal—still in the conceptual stages—calls for 853-foot-tall (260 m) turbines that can each generate enough electricity to power 16,000 American homes. This turbine, called the Haliade-X, is already in use in shallow offshore farms, but in the new installation, towering structures float via three-legged tension-leg platforms similar to those supporting offshore oil or gas rigs, with so-called active tendons that secure the platform to the seafloor and help it ride the waves.

An offshore wind farm in Germany

are growing endeavors to introduce mass-scale wind and solar farms around the world. Today, about a third of our global electricity comes from renewables; the United Nations projects that portion could hit 90 percent by 2050.

In the United States, wide adoption of wind and solar will require funneling energy from those sources into cities, where Americans consume most of their energy. That's a challenge. Current data show the wide discrepancy between midwestern states with lots of space for wind and solar farms but not as much demand, and population-dense states with more urban sprawl, less space for energy farms, but much higher needs. Iowa received more than 60 percent of its power from wind farms in 2022, and South Dakota, Kansas, and Oklahoma all cleared 40 percent. By contrast, New York generated just shy of 4 percent of its energy needs from wind (though the Empire State performs better in other categories, particularly hydroelectric generation).

While many engineers puzzle over how to transfer all this energy to where people need it, others have turned their ambitions elsewhere, to fantastic forms of energy that seem to only exist in science fiction. Yet these visionary engineers have made tremendous strides that could lead to breakthroughs that leapfrog wind and solar entirely. The future of energy could be an evolved version of our current picture, or it could come from a source we can't even harness yet. That's what we're exploring next.

A German engineer operates on an offshore windmill 300 feet (91 m) in the air and 30 miles (48 km) from the mainland.

4

The Search for a Clean Energy Future

T **THE ENERGY SOURCES** we've discussed up to this point are all either in use or well along in the development process. But as the world searches for a path through what is sometimes called the Great Conversion—the journey from fossil fuels to green renewables—we have put many more possible energy sources on our drawing boards.

Some of the science community's attention is centered on new sources of energy, like nuclear fusion. Others are looking for inventive ways to tap old energy sources, like linking wind farms to hydroelectric systems (as at Hoover Dam), or satellites that collect solar power. Yet others are pursuing carbon neutrality from the other

An engineer conducts maintenance inside the Joint European Torus facility's tokamak, a doughnut-shaped chamber lined with magnets.

direction and trying to suck carbon dioxide from the atmosphere as it's produced. All these efforts to minimize climate change are like threads in a complex tapestry we are weaving to change our way of life in the world.

NUCLEAR FUSION

THE LIVERMORE FUSION research facility represents just one of two ways humans can pull energy from the nucleus of an atom: It's a laser fusion laboratory, still in the testing stages, whereas many fission reactors are already up and running today. Both fusion and fission involve reactions whose end products have less mass than the initial ones. The difference in mass, via Einstein's famous equation $(E=mc^2)$, is converted into energy.

Fission involves splitting a large nucleus, like that of uranium (atomic number 92), into smaller pieces. This type of reaction drives the nuclear reactors already producing about 10 percent of the world's electricity, according to the International Energy Agency. They'll certainly be around in the future—the science behind the reactors is solid—but their proliferation will depend more on political and economic decisions. Mining uranium, nuclear waste treatment, safety concerns—all are complex issues involving many parties.

Fusion, on the other hand, may answer some of these concerns. It involves combining two small nuclei to create a larger one (along with a spray of particles). The most familiar fusion reaction (combining four hydrogen nuclei to make one helium nucleus) is what powers the sun. If we could learn to control fusion reactions, we would essentially have an inexhaustible source of energy, because the world's oceans are full of hydrogen-rich water. Would that drain the oceans? Not even close. A physicist at UC Davis calculated that a single shower's worth of water would be enough to meet a person's energy needs for an entire year. The sun will die before we could use up the world's oceans.

▶ The Obstacle to Fusion

THE FUNDAMENTAL PROBLEM blocking the development of fusion power is simple. To get two protons (the nuclei of the hydrogen atom) close enough for nuclear reactions to take over, the protons have to be moving

A technician inspects the interior of the Target Chamber at Livermore's National Ignition
Facility, where scientists are testing light-induced nuclear fusion.

A colorized image of a fusion implosion: The cylinder at the center of the lasers' blasts measures 5.75 millimeters in diameter.

This two-millimeter target capsule for fusion experiments is filled with cryogenic hydrogen fuel. The capsules are made of plastic, diamond, or beryllium.

fast enough to overcome the electrical repulsion between them, because they are both positively charged. This requires a high temperature. (The sun's core, for example, measures about 27,000,000°F, or 15,000,000°C.) In other words, to establish an energy-producing fusion reaction, you have to create a hydrogen gas with a temperature of millions of degrees.

And what do you put that hot gas into? The candidates we might think of immediately—steel or concrete boxes—would instantaneously vaporize. At the moment, there are two high-tech attempts to deal with this fundamental question. One approach is magnetic confinement fusion (see sidebar on page 76), and the other is to trigger a fusion reaction by setting up shock waves in an isolated drop of solid hydrogen.

▶ Fusion Through Lasers

ANOTHER PATH TO FUSION is called inertial confinement, the technique deployed at the National Ignition Facility at Lawrence Livermore National Laboratory. The most common way to achieve this type of fusion is by blasting a drop of frozen hydrogen with powerful lasers—in Livermore's case, 192 such lasers.

Generally, in this method, the frozen hydrogen is actually a mixture of two different isotopes of hydrogen: one called deuterium (consisting of one neutron and one proton) and another called tritium (two neutrons and one proton). When the lasers blast the mixture, the outer layer boils off, creating shock waves that raise the temperature of the drop high enough to initiate fusion in the center of it. But that's not the last step. To achieve ignition (that is, to generate energy), the helium nuclei produced by fusion have to heat up the outer layers of the drop and initiate fusion there as well. In essence, ignition is achieved when the fusion reactions in the frozen drop produce more energy than was carried in by the lasers.

In 2022, the Livermore lab researchers showed they could initiate fusion. And they've repeated that accomplishment seven more times as of this writing. This is a landmark achievement in science, but unfortunately, it does not mean the system could be used for power generation any time soon. The energy required to operate the massive laser system still exceeds the energy produced by fusion, though further gains could tip the scales in the other direction.

▶ Magnetic Confinement Fusion

THE FUSION WORLD has centered a large portion of its attention (and money) on the so-called ITER system being built in southern France. (ITER means "forward" or "the path" in Latin, but it originally stood for International Thermonuclear Experimental Reactor.) This system, the result of collaboration of 35 countries including the United States, is an example of magnetic confinement fusion, which uses a magnet-laden reactor known as a tokamak to create a fusion reaction inside a hollow, doughnut-shaped container (see sidebar below).

FROM SOUP TO FUSION

A At the high temperatures needed to maintain fusion reactions, hydrogen gas becomes a plasma—a soup of disconnected nuclei and electrons. If that plasma is in a magnetic field, then the plasma locks on to the field—move the field and you move the plasma and vice versa.

That's the basic principle behind magnetic confinement fusion. As the plasma is heated, the field is pulled away from the walls of the container. In the end, the plasma never touches anything, but is held in a vacuum by the field. Thus, magnetic confinement fusion allows us to hold hot plasma for long enough that deuterium and tritium can overcome electrostatic repulsion and fuse via strong nuclear force. Most experimental fusion reactors use this kind of system—except, notably, Livermore.

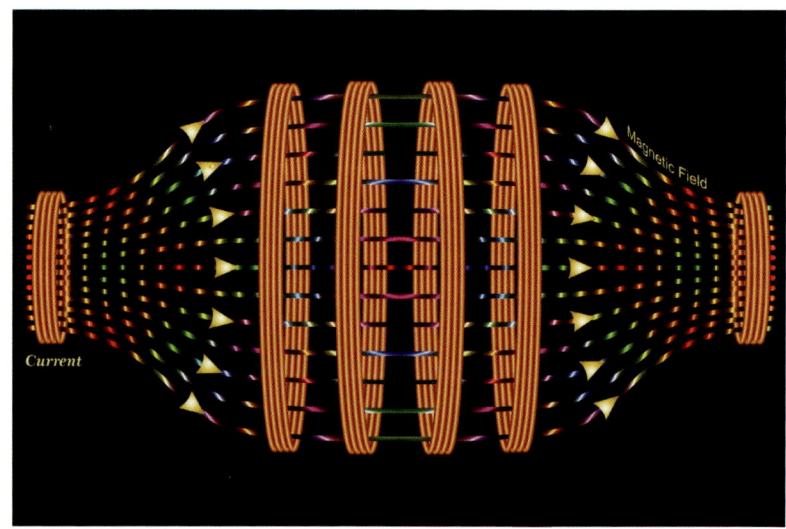

Magnetic Field

Current

This diagram displays how plasma particles can be suspended in a magnetic field.

Spokespeople for ITER have explained that the project is meant to demonstrate the possibility and replicability of controlled fusion, laying a template of sorts for future reactors. Basically, this is the first step toward commercial fusion power. But ITER has faced numerous delays and ballooning budgets. Under its initial 10-year plan, it was due to come online in 2016 at a $6.3 billion budget. Then in an update, it was estimated to come online in 2025 at a $22 billion budget. Most recently, spokespeople for ITER have suggested a target date of 2035, with an undisclosed price tag.

Many physicists are cynical about nuclear fusion power for a reason. A standing joke says that fusion is the energy source of the future and always will be. For now, scientists are making incremental progress that is serving to ground the technology in some realm of feasibility. As we chip away at that mother of all scientific problems, though, other innovations in well-tested forms of energy are evolving our energy landscape.

SOLAR POWER SATELLITES

SOLAR ENERGY IS BOTH renewable and easy to harvest, but our current collection methods raise a few disadvantages, including the intermittence of weather, the diffuse nature of solar energy when it reaches Earth, the day/night cycle, and the need for vast open spaces. But since the 1970s, a small group of scientists and engineers have looked at these problems in a novel way. Why, they ask, do solar collectors have to be located on the surface of Earth? What if we put the collectors in space, above Earth's atmosphere, then beam the power down to where it is needed?

We might be closer to building a system like this than you think. As of this writing, Japanese scientists are planning to test a miniature space-based solar power plant in low Earth orbit. The 400-pound (180 kg) satellite will, in theory, transmit 1 kilowatt of collected solar power via microwaves toward an antenna on Earth. That's only enough power to run a household dishwasher for a single cycle, but hey, it's a start!

▶ The Pros and Cons of Building Solar Power Satellites

COLLECTING SOLAR ENERGY before it enters Earth's atmosphere has a lot of theoretical advantages. Clouds, seasons, and nighttime darkness would not affect a receiver in space, especially if the solar collectors were placed in geosynchronous orbit (meaning their orbital period matches Earth's rotation period). In this case, the collector would be 22,000 miles (35,000 km) above Earth's Equator and, from the point of view of someone on the planet's surface, would be stationary in the sky. In that position, the collector would be in perpetual sunshine.

The pros are obvious, but we have to confront two fundamental questions: How can we build these huge collectors in space, and, once the solar energy we want is collected, how can we send it back to Earth? We will have to figure out how to cover a few square miles of space with enough solar collectors to generate a justifiable amount of energy, an amount that makes building the satellites worthwhile in the first place. Needless to say, this kind of construction is beyond our abilities at this time, but let's say we launch our solar-collecting satellites. Engineers have proposed solutions to getting harvested energy back to Earth that involve using microwave beams (or lasers) to send signals to a receiver on the ground. To keep the beams from harming people who might wander into them (!), conceptual designs call for receiving stations with an area of 10 square kilometers (3.86 sq mi) or more to dilute the beams' power. Whether such a system could actually transmit gigawatts of power remains an open question.

| CARBON CAPTURE AND SEQUESTRATION |

MANY OF THESE RENEWABLE energy ideas are feasible, but will take a long time to develop or implement in a way that makes a dent in our current reliance on fossil fuels. In the meantime, scientists are exploring how to remove carbon already in our atmosphere. This process is called carbon capture, and many methods are already in use today.

▶ Smokestack Scrubbers

SOME SYSTEMS THAT use fossil fuels to generate electricity have installed various types of smokestack scrubbers, well-known but expensive ways

A drone's-eye view of an oil and gas jack-up, a rig that elevates drilling equipment using legs anchored to the seafloor

to remove carbon dioxide from their emissions before it is released into the atmosphere. These include wet scrubbing, in which a solvent (like water) washes pollutants from a gas stream, and dry scrubbing, which achieves the same adsorption through a gas solvent instead of a liquid. In either case, certain substances in the compounds "target" pollutants and carry them away from the emissions.

▶ Direct Air Capture

ONCE CARBON DIOXIDE is in the atmosphere, it's too late to scrub the pollutant, but we can still pull the air into chemical reaction chambers with giant fans and remove carbon dioxide from the nearby air. This direct air capture technique uses the same general principle as smokestack scrubbing—a chemical reaction washes carbon dioxide (CO_2) from the air—but just in a different place. The main problem with this approach at the moment is that, although it is relatively easy to make carbon dioxide bind to various chemicals, freeing up those chemicals to repeat the capture process economically is difficult.

▶ Silicate Weathering

OVER GEOLOGICAL TIME spans, carbon in the atmosphere is incorporated into materials like limestone and held there by atomic forces. In this complex natural process, carbonic acid made from rainwater and CO_2 dissolves silicate rock to make ions. Scientists are searching for ways to speed up this process to use it for deliberate carbon sequestration. The technology is still in the experimental stages, but if the research bears fruit, carbon removed from the atmosphere will be stored in mineral form, and can be trapped there on any time scale we wish.

▶ Geoengineering

ALL THESE METHODS described so far are based on the theory that the easiest way to lower the amount of carbon in the atmosphere is to refrain from putting it there in the first place. But that doesn't mean that in the long-term, unknown future, we should dismiss other measures out of hand.

The most controversial suggestions ever made concerning the future of the planet and climate change involve geoengineering. The general idea behind this concept is this: If human behavior is changing Earth in ways we don't like, then we ought to modify Earth so it better accommodates human behavior. Applied to the problem of carbon dioxide in the atmosphere, geoengineering techniques are being designed as attempts to change the amount of carbon dioxide in the atmosphere, by either removing it or preventing its arrival.

TINY NUCLEAR REACTORS

R Right now, most nuclear fission reactors, towering behemoths that dominate the outskirts of urban centers, operate at the gigawatt scale. But experts say nuclear technologies could grow their share of the energy picture by combining the giants with dwarfs and developing tiny nuclear reactors on a megawatt scale.

A tiny nuclear reactor like the one designed by Portland-based company NuScale could power 650 homes with a reactor the size of a school bus. That's 10 percent of a plant's capacity in just one percent of the space, making a tiny reactor ideal for small towns and remote corners of the grid. Engineers claim tiny reactors are safer than power plants too. Many designs use physics-dependent risk management systems instead of the human- or machine-dependent safeguards at power plants. So if something goes wrong, you're relying on gravity, not Homer Simpson, to save the day. The tiny-reactor concept is gaining traction: In 2025, the U.S. Nuclear Regulatory Commission certified NuScale's 77 megawatts electric (Mwe) version of this reactor design.

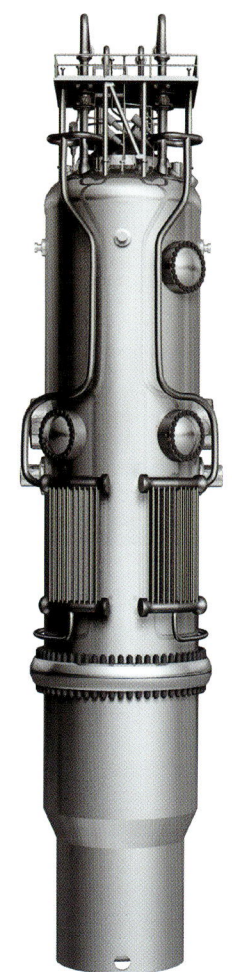

NuScale Power's tiny reactor module is the size of a school bus.

▶ Solar Radiation Management

ONE SUCH TECHNIQUE is called solar radiation management (SRM). The temperature of Earth depends on the amount of sunlight that reaches and warms its surface. With SRM, material designed to reflect sunlight back into space before it reaches Earth is put into the atmosphere or in orbit, thus lowering the planet's temperature.

Fortunately, nature has already shown us a way to do this. In 1816, the world experienced a horrendous year without a summer. The global temperature fell by about 1°F (0.56°C), widespread crop failure ensued, and many people died as a result of the food scarcity. We now know that the cause of this disaster was the eruption of Mount Tambora, a volcano in what is now Indonesia. The explosion threw tons of sulfate particles and other substances into the atmosphere, and for months on end, this material reflected incoming sunlight back into space.

That tragic year inspires futuristic possibilities. The basic idea behind the most popular version of SRM is to inject sulfate particles into the atmosphere—in effect producing an artificial volcanic eruption. There have been serious engineering studies about how such a system would work, but researchers point to some challenges. The sulfate particles have to be injected from specially designed aircraft flying at an altitude of about 65,000 feet (20,000 m), a bit above the altitudes commercial aircraft normally use. A serious program would require the design of specialized aircraft, but that's possible with current technology. To counter current carbon emissions, a fleet of about 100 aircraft would need to eventually make about 60,000 flights a year, with a cost of about a billion dollars a year worldwide.

▶ Plankton to the Rescue!

ANOTHER GEOENGINEERING strategy focuses on the oceans: Plankton to the rescue! Scientists estimate that the oceans absorb about 30 percent of the carbon dioxide emitted by human activity. You can, in fact, think of the oceans as something like a giant carbonated soft drink. Moreover, the oceans provide a path to sequestration of the carbon they absorb: At the ocean surface, any one of a number of species of plankton extract carbon from the water and use it to construct their microscopic skeletons. When the plankton die, these skeletons (along with the carbon they contain) sink to the ocean floor, where they are eventually incorporated into limestone. Thus, the carbon is permanently removed from the atmosphere.

So how can we foster these plankton eco-warriors? Scientists have been exploring how to fertilize the ocean to foster plankton growth without disrupting the broader ecosystem. In a first known attempt in 1993, scientists added iron to areas of the Southern Ocean. Those on the expedition reported that the added iron turned the water into a green goo as plankton reproduction went into high gear, but a later study showed the iron fertilization might have contributed to native fish losing biomass.

Today, scientists who advocate for geoengineering stress the importance of paying attention to possible unintended side effects on existing ecosystems. In addition, the relatively low cost of these interventions has caused political scientists to consider the political difficulties that geoengineering might trigger. What if country A started an SRM program that benefited their farmers, but hurt farmers in country B? It might ignite a climate war, and clearly, that's the wrong battle to fight.

HOW WILL WE POWER THE FUTURE?

THE SEARCH FOR THE holy grail of energy is a little-by-little campaign, one that requires threading tight needles and walking fine lines. Unfortunately, we're following this course behind the wheel of a massive cruise ship. We are working against precedent, legislation, current infrastructure, geography, cultural norms, and miles of red tape on the way to cleaning things up and putting wind, solar, nuclear, and other energy innovations across larger parts of the map.

And as the debates swirl, our demand grows. Technology is more attractive, helpful, and productive than ever, so our reliance on devices, lighting, automation, and such is going up, up, up. That requires a lot of energy—and specifically, a lot of electricity.

Though competing industries jockey for dominance in energy production, electricity is finding its way into everything. In the past century, what was once an eyebrow-raising realm of science has become the central power source in our homes, the emerging power source on our roads, and ... as we'll see, a trailblazing source in the skies. Our energy picture is tied closely to our electricity picture, so that's where we're going next.

The Solúcar solar farm in Spain

II
ELECTRICITY

Flying the Aircraft That Will Shock the Skies

AS THE TIP OF A GOLDEN SUN peeks over the central Florida horizon, something buzzes through the air in front of me—the largest drone I've ever seen, large enough to carry a person. In fact, its 18 spinning propellers are currently supporting a professional pilot from LIFT, a start-up company taking personal air transport into the future. This drone—a single-seat, ultralight aircraft named Hexa—is classified as an eVTOL, or electric vertical takeoff and landing. As the clock strikes 7:00 a.m., the opening music for *Good Morning America* chimes through my earpiece. Then I hear the voice of Michael Strahan:

"George Jetson is live … I mean, Rob Marciano is live! Ready to master the game of drones and take the first-of-its-kind aircraft for a spin …"

That's right. Strahan is teasing my segment coming later in the show, when millions of viewers will watch me fly the LIFT Hexa myself! It's an exciting moment for me, and although LIFT gave me an hour of classroom and simulator training, my moment on the air will be my first real flight. So yes, I'm nervous.

Before I take off, the *GMA* control room rolls a piece we taped on the burgeoning industry of eVTOLs—the latest innovation in the private-sector race toward electric flight. Like helicopters, eVTOLs can take off vertically, transition to horizontal flight, then land vertically, and they can do it just

Archer Aviation's Midnight, an electric air taxi, has a 60-mile (100 km) range and a maximum speed of 150 miles an hour (240 km/h). The aircraft began advancing through a rigorous flight-testing program involving multiple phases in 2024.

about anywhere, including your backyard. Because they run on electric power, they operate with nearly zero noise, pollution, or carbon emissions.

That's a big deal. Personal transportation—bikes, cars, buses, trains, whatever takes our tushes from one place to another—might be the quintessential driver of modern society. While most modes of transportation are already going electric, fixed-wing and rotary aircraft are behind the pack. The aviation industry accounts for 2.5 percent of global CO_2 emissions, and private cars and vans account for around 10 percent. The eVTOLs are the first step in cutting back on these percentages. They're nimble enough to take off and land in tight spaces—a must in densely populated areas like Manhattan—and accessible enough financially and operationally to make electric flight ubiquitous. The leaders in the eVTOL space say a ride in one of their aircraft shouldn't cost more than an Uber.

These promises might sound crazy, even dangerous. As I prepare to fly Hexa, I recall the data showing single-engine aircraft have proportionately more accidents than other types of aircraft. I also recall the words of my father, who was a military and airline pilot. Whenever I boarded a news helicopter, he'd always tell me: "If that one engine fails, you're likely dead." *Gulp.* Thanks, Dad. Good chat! Fortunately, I think my dad would feel better about eVTOLs. There's little risk of electric aircraft failing mid-flight because they often use several small motors instead of a single large one: In fact, Hexa has a motor for each of its 18 propellers. If one electric motor goes kaput, the others will keep on humming.

I remind myself of this as I climb into the cockpit. LIFT's chief pilot has already powered up the propellers, and he gives me control. I slide the goggles off my helmet and over my eyes, then reach for the joystick on my right. It has three buttons: TAKEOFF, LANDING, and HOME. Two toggles flank the stick, one on each side. My thumb rests on an up/down toggle, while my index finger wraps around to a toggle that controls yaw—part of how an aircraft turns. When LIFT gives me the signal, I count out: "Three ... two ... one ... taking off," and press the TAKEOFF button as I leverage my thumb on the up toggle. Gravity presses my butt to the seat as Hexa launches into the air. I can't help letting out a thrilled "Whaaaaooooo!" as my ride into the future begins.

Hexa is surprisingly simple to fly: There's just one joystick and a tablet above my knees that displays easy-to-see instrumentation. Meanwhile,

LIFT's Hexa aircraft in 2019: The same craft took author Rob Marciano to the air and landed again safely live on *Good Morning America.*

Archer Aviation built a manufacturing facility in Covington, Georgia, in 2024. The company hopes to scale production here to 650 aircraft annually by 2030.

my left hand has nothing to do, and my feet have no pedals to manage. The barrier of entry for this eVTOL is low. Just about anyone could fly it.

For engineers, electric motors offer tons of benefits. They are smaller and lighter than combustion engines, easier to maintain and simpler to test thanks to fewer parts, quieter to operate, and they don't require messy, flammable lubricants. Later in this section, we'll dissect how an electric car works, and you'll see that the tech is easily translatable to an airplane. We've seen massive growth in electric cars, and it seems electric aircraft are ready to follow.

My two-minute demonstration flight wraps up with another cue from Michael Strahan: "All right, Rob, you think you can stick the landing?"

By this point, I'm confident. I reply: "I got this, Michael." But in reality, Hexa has it. All I have to do is press the HOME button on the joystick and Hexa flies itself over the landing zone. Then, I press the LANDING button and the aircraft descends.

Everyone in the studio is impressed with my landing skills until I let go of the joystick and extend my arms. "Look, Ma, no hands!" I shout. The entire sequence is on autopilot, and the aircraft indeed sticks the landing. It's a heck of a kicker for this amazing innovation.

LIFT is not alone in the eVTOL industry, soon to command well over $20 billion. The Boeing-backed company Wisk hopes to have an autonomous four-seat eVTOL to market as an air taxi—it would function with an app, like Uber—and United Airlines has contracted 200 of

Bell's Nexus 4EX aircraft can convert from helicopter mode to airplane mode, as shown at a 2020 exhibition here.

Archer Aviation's Midnight, a five-seat, all-electric passenger aircraft, could one day be used for personal transport.

Archer Aviation's five-passenger Midnight eVTOLs for quick commutes to and from major airports. In a Midnight, you could hop from Newark International Airport to Manhattan in mere minutes.

I paid a visit to Archer Aviation's headquarters and design facility in San Jose, California, where I met Adam Goldstein and Tom Muniz. Goldstein is Archer's founder, while Muniz is chief technology officer and one of the company's lead engineers. Like many Silicon Valley start-up guys, they are relatively young, casual, and exuberant about what they're building. "It's not really an engineering challenge anymore," Goldstein says about building eVTOLs. "Getting it through the regulators is now the biggest challenge." These regulators would certify the aircrafts and, at some point, redesign the airspace to create streets in the sky, essentially. With backers like Boeing, United Airlines, and the U.S. military on board, it looks like the aviation industry is betting that eVTOLs will clear the red tape. Archer is counting on it. As of this writing, they plan to have a fleet of Midnight aircraft operating in Los Angeles before, during, and after the 2028 Olympic Games.

For their part, LIFT spokespeople conceded to me that the single-seat Hexa I flew will initially target thrill riders on a pay-per-flight basis. Later, Hexas could potentially be sold for use on large chunks of private or government-owned land. In the far future, it could be the archetype for personal eVTOLs. That has me wide-eyed, the thought of buzzing around in my very own eVTOL like George Jetson, running errands and zipping the kids to school … things are starting to look like the future I saw on television when I was a kid!

The promises of an electric future are alluring because the basic science of electricity seems in and of itself too good to be true. Here's an energy source that's readily accessible, abundant beyond measure, and exceptionally efficient. It *should* be everywhere! Across these next four chapters, we'll venture inside the tiny but mighty electron and learn how it makes the juice powering nearly all of modern life. We'll learn how the grid keeps the lights on, what goes into the world's longest-lasting battery, and what we need to support a society full of electric cars and aircraft. The next big thing isn't far away. In fact, thanks to our old friend electricity, it's already in our homes, on our roads, and flying over our heads!

The electric flying car company XPENG conducted the first flight of its X2 aircraft, complete with gull-wing doors, in 2022.

Electric air taxis, such as the VoloCity helicopter shown here, are still trying to take off. Diamond Aircraft acquired Volocopter in 2025.

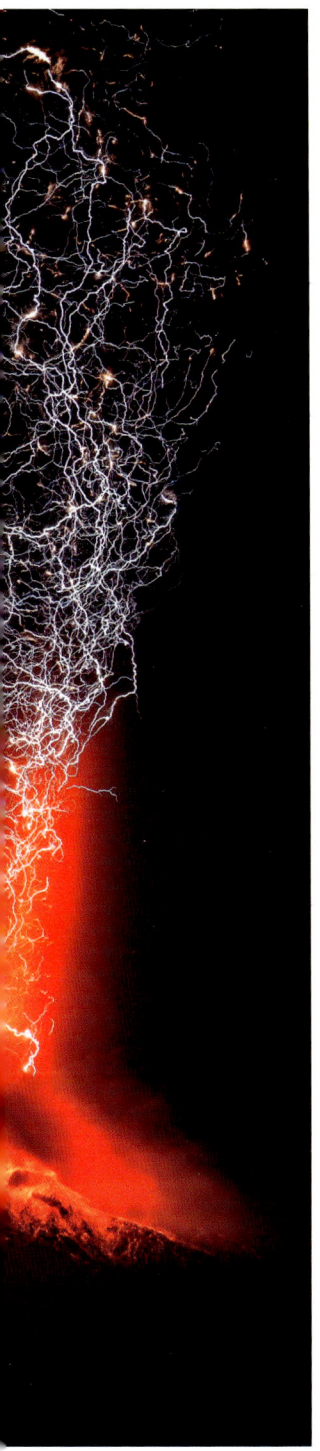

The billowing ash plume from the 2015 Calbuco volcano eruption in southern Chile generates a dramatic electrical storm.

CHAPTER

5

Explaining Electricity

E

ELECTRICITY IS AN amazingly convenient form of energy. It can be generated in one location, transported hundreds of miles, then used somewhere else. Regardless of the amount of pollution produced in the generation process, at the point of use, electricity is as clean a source of energy as there is. This makes it ideal for use in cities, where 56 percent of all human beings live now. Those of us not already driving an electric car likely will be soon. And who knows—someday we may all be flying in electric aircraft like a Midnight, or even in an electric passenger plane. Most of the world's advanced economies are engaged in a massive conversion to energy systems in which electricity will be the primary energy source. The future is looking more electric by the day.

Most electric technologies won't look much different from similar technologies running on another energy source. After all, from the point of view of the driver, an electric vehicle isn't that different from one running on gasoline. But what will be different is how we'll generate electricity—and that will alter how the future looks, from widespread wind turbines and solar panels to electric motors in just about everything. With this in mind, it is important to understand what electrical energy is and how it works.

| ELECTRICITY AND MAGNETISM |

WHENEVER A SOCK hitches a ride on a towel on its way out of the dryer, it's exhibiting a perfect example of one of the fundamental forces that make the universe function. While the sock and towel were tumbling in the dryer, electrons were scraped off the sock and deposited on the towel, creating a force we call static electricity. Friction pulls electrons away from the atoms in one material (in this case, the sock), leaving those atoms and that material with a positive electrical charge. At the same time, friction deposits those atoms in another material (in this case, the towel), leaving the second material negatively charged. Just doing the laundry, then, reveals a fundamental law of nature: There are two kinds of electric charge—positive and negative—and opposite charges attract each other, while like charges repel each other. This statement is known as Coulomb's law, after the French scientist Charles-Augustin de Coulomb (1736–1806), who first formulated it.

The experiments continue in the kitchen. Most refrigerator doors are plastered with all sorts of stuff: souvenir magnets from various travels, magnets holding up a shopping list, family photos turned into magnets, and so on. These magnets stay on the refrigerator door despite the fact that Earth is trying to pull them down with the force of gravity. Obviously, we have found another fundamental force in nature—magnetism.

Because Earth itself is a giant magnet, people learned to use this fundamental force early on as a navigation aid, and scientists worked out the basic rules that govern the behavior of magnets. An ordinary bar magnet has two ends. These are called poles, customarily labeled north and south. The rule that governs magnetism is: Every magnet must have at least two poles. Like poles repel each other; unlike poles attract.

The reason that some materials, like iron and nickel, have magnetic properties whereas other materials do not has to do with forces operating at the atomic level.

▶ Magnetic Fields

ONE WAY OF VISUALIZING the magnetic force is to picture a magnetic field with the help of a compass needle: Lay down a compass at any point near

▶ HOW COMPASSES KNOW DIRECTIONS ◀

Humans have used compasses since at least 200 B.C.E., when the Chinese used early versions of the compass to guide them on their travels. Given that our planet acts like a giant magnet, any smaller magnet laid down on Earth's surface will be subjected to a magnetic force, opposites attracting. The magnet's south pole will be attracted to Earth's north pole in northern Canada, while the magnet's north pole will be attracted to Earth's south pole in Antarctica. Thus, the magnet (in the form of a compass) will line up in a north-south direction, allowing us to determine directions.

A technical point of possible confusion: Though the south pole of the compass magnet points north, this pole is customarily labeled with an *N* solely because it points north. Thus, the south pole of the compass magnet is actually called the north-seeking pole.

Analog compasses can be more accurate than apps, which are prone to interference.

a magnet, and the direction the needle points will indicate the direction of the magnetic field. Therefore, laying down multiple compasses and aligning their needles nose to tail would create a continuous line that represents the magnetic field's complete orientation. When a magnet has two poles, its magnetic field is called a dipole field. The compass needles in a dipole field would connect one pole to the other in tight arcs.

▶ Earth's Magnetic Field

EARTH'S MAGNETIC POLES exist in part because of the rotation of Earth's liquid outer core. Earth has its own surrounding magnetic field as well,

▶ VOLTA, GALVANI, AND DR. FRANKENSTEIN ◀

The development of the battery grew out of a debate about the role of electricity in living systems. The Italian physicist Luigi Galvani (1737–1798) showed that applying electric current to the legs of newly dead frogs produced muscle contractions. He argued that living things possess a vital force he called animal electricity. Volta, on the other hand, argued that chemical reactions in the frog's legs triggered electrical signals—a view that we accept today. Volta actually developed the world's first battery, the voltaic pile, to disprove Galvani.

Unfortunately, popular imagination ran against Volta's experiments in unexpected ways. As the debate over animal electricity grew, public exhibitions about the effects of electricity proliferated. These frequently involved demonstrations in which human corpses would be jolted with electric current, and urban legend holds that reports of these shows inspired Mary Shelley as she was writing her novel *Frankenstein*.

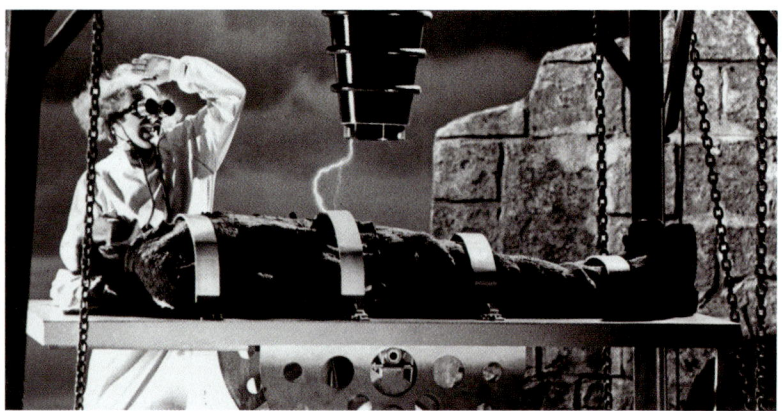

"It's alive!" thanks to electricity: a still from Mel Brooks's *Young Frankenstein* (1974)

which triggered early human interest in magnetism. The shape of that field is, remarkably, the same shape as that of an ordinary dipole magnet. In fact, Earth's magnetic field deflects a stream of charged particles (called solar wind) that the sun is constantly emitting, thus protecting the living things on our planet. Perhaps the most dramatic effect of the planet's magnetic field are auroras. When electrified particles from the sun are bent toward Earth's poles by our planet's magnetic field, the particles interact with gases in the atmosphere to produce auroras. Scientists can study this phenomenon to learn about the inner working of the sun, and what the sun is ejecting into space.

BATTERIES AND ELECTROMAGNETISM

WE OFTEN SEE THE LAWS of magnetism and electricity in our daily lives through batteries. Each battery inside our phones, TV remotes, and cars operates on the same basic principles—and they all originated from a single unlikely invention. In 1800, the Italian physicist Alessandro Volta (1745–1827) invented something he called a voltaic pile—what we call a battery today. Made of alternating disks of two different metals separated by paper or cloth and sitting in salt water, this device reliably converted chemical energy into electric current, as all batteries now do. Volta's invention marked the first time scientists could study the effect of moving charges, or currents, as opposed to static charges.

In 1820, the Danish physicist Hans Christian Ørsted (1777–1851) used Volta's device to obtain the first experimental results that indicated there might be more to the story of electricity and magnetism than the results we have described for static systems. According to a popular legend, he stumbled on his discovery accidentally during a lecture. He had one of Volta's batteries on a demonstration table, and right next to it was a large compass. He noticed that whenever he turned on the current for Volta's battery, the compass needle would swing around. This could only mean that the current was producing a magnetic field. Electricity and magnetism were not as separate as scientists had assumed. Ørsted's result demonstrated another law: Moving electrical charges produce magnetic fields.

▶ Electromagnets

ØRSTED'S DISCOVERY IS the basis for a common device known as an electromagnet. If a current-carrying wire is wrapped into coils, it produces a magnetic field identical to the field produced by a bar magnet (see diagram below). Thus, we have a magnet that can be controlled by an electric current—and the more current we feed into the magnet, the stronger the magnetic field becomes. The magnet can even be turned on and off with the flick of a switch.

The giant magnets used to move piles of scrap metal are an intuitive example of electromagnets in action. It's easy to understand how these magnets lift the metal—the magnetic force generated by the electromagnet counteracts the downward pull of gravity. But how does the magnet let go of the metal once it has gone into operation? With the knowledge of how an electromagnet works, the answer is obvious—the operator simply turns off the current to the magnet, the magnetic force disappears, gravity takes over, and the metal falls.

The fact that electrical currents can produce magnetic fields explains why they always, to our current knowledge, come in pairs. Think about the exercise of cutting a bar magnet in half, then cutting a half in half, and so on. Eventually, this process will produce a single atom. This atom will have electrons—charged particles—in orbit. Those orbiting electrons actually constitute an electric current, and so, according to what Ørsted found, they will produce a magnetic field. In this way of looking at things, magnetism at the atomic level is a consequence of the structure of the atom. Take the atom apart and that structural magnetism disappears.

ELECTROMAGNET

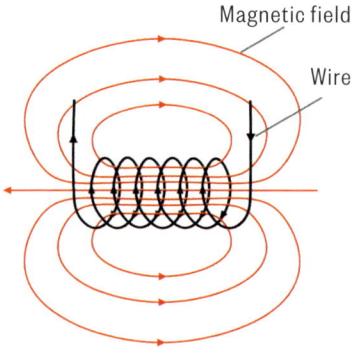

Magnetic field

Wire

Yet scientists have also found that the electrons in orbit and the particles in the nucleus actually generate their own magnetic fields. They are in essence subatomic bar magnets. And if we took these particles apart (presumably, in the case of electrons), we would find another level of structural magnetism similar to what we find in the atom.

An electromagnet lifts a pile of iron filaments. With the flip of a switch, the magnetic field can be turned off, releasing the iron.

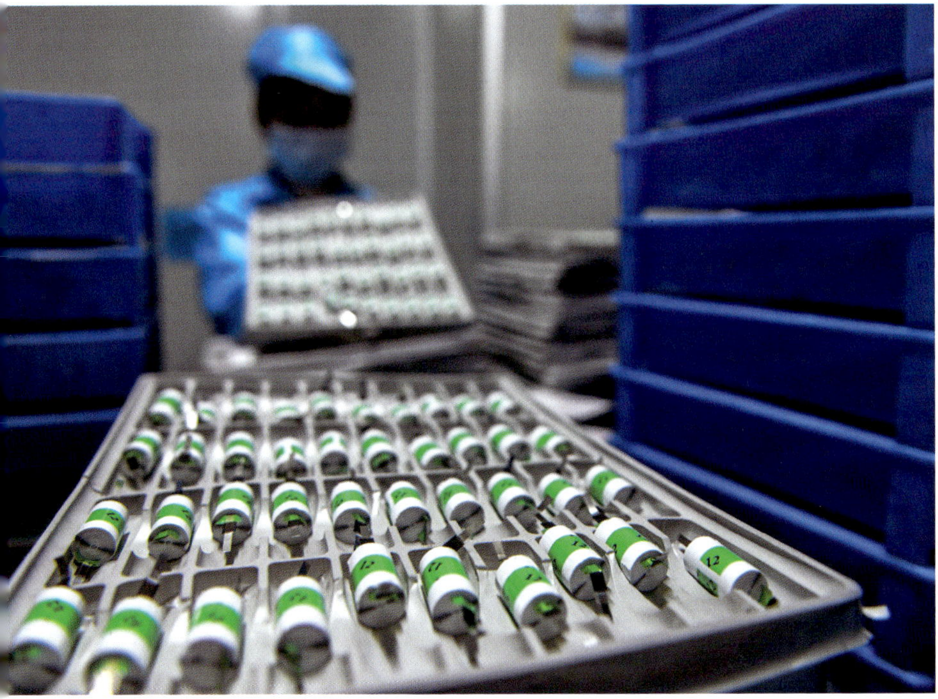

A lithium battery production line in Yongzhou, China, demonstrates one of the future's most critical areas of development and commerce.

Scottish physicist James Clerk Maxwell connected four equations that make up what we know today as the four basic laws of electricity and magnetism.

▶ A Bookbinder's Hidden Genius

THE NEXT STEP IN ESTABLISHING the connection between electricity and magnetism was taken by English scientist Michael Faraday (1791–1867). Faraday was the son of a blacksmith and grew up in the Sandemanian church, a sect of Presbyterianism. This meant that even in the unlikely event that his family could send him to university, he would have been denied admission because he was not a member of the Church of England. Consequently, he apprenticed with a bookbinder to learn a trade. As he read the books that came into his shop, Faraday cultivated a particular interest in science. He attended a series of public lectures given by Sir Humphry Davy, the most prominent scientist of the time. After taking notes in a beautiful early 19th-century script, he bound them in leather and presented them to Davy as a calling card. Davy hired him as a lab assistant (who wouldn't have?). Thus, despite coming from humble beginnings, Michael Faraday became the greatest scientist of his time.

Faraday's most important discovery was a process known as electromagnetic induction—the process we use to generate most of the electricity we use today. Just as Ørsted showed that moving electrical charges can produce magnetic effects, Faraday showed that changing magnetic fields can produce electrical effects.

Here's an easy way to visualize the process of electromagnetic induction: Imagine a loop of wire lying on a table, not connected to a battery or any possible source of electrical current. Imagine a magnet suspended above that loop, and one of the following happens: (1) The loop moves, but the magnet doesn't; (2) the magnet moves, but the loop doesn't; or (3) both the magnet and the loop move. Faraday found that a current will flow in the loop in all three cases. In other words, whenever the magnetic field around the loop changes, either because we've changed the position of the magnet or the position of the loop, a current will flow in the wire. This is the basic principle behind the operation of the electric generator.

A favorite Faraday story (though one that's likely apocryphal) concerns the day the prime minister of Britain, Lord Palmerston, was getting a tour of Faraday's laboratory and was shown a prototype of the electric generator. The following dialogue ensued:

"Very interesting, Mr. Faraday, but what good is it?"

"What good is it? Why, Mr. Prime Minister, someday you'll be able to tax it!"

▶ Maxwell's Equations

OUR EXCURSION THROUGH the phenomena of electricity and magnetism has turned up four basic laws: (1) Like electric charges repel each other; unlike charges attract. (2) A magnet must have at least two poles. Like poles repel each other; unlike poles attract. (3) Moving electrical charges create magnetic fields. (4) Changing a magnetic field creates electrical current.

These equations, grouped together, are knows as Maxwell's equations, after the Scottish physicist James Clerk Maxwell (1831–1879). He was the first person to see that these equations form an interconnected whole, and he had the mathematical ability to take advantage of that insight.

The floppy disk motor and circuit board shown here are the guts of systems that most readers born after 1990 have likely never seen before!

The Large Hadron Collider, a particle accelerator, consists of a 16.5-mile (26.6 km) ring of superconducting magnets.

Overemphasizing the role of Maxwell's equations in modern science is impossible. Every phenomenon involving electricity and magnetism—from binding an atom into our DNA to the motion of gas clouds in space to developing the eVTOL Midnight—is contained within them. Furthermore, when applied to the behavior of elementary particles and used to predict the result of experiments, they are one of the best-verified theories in all of science. It is as if we had a theory that could predict (for example) the distance between New York and Los Angeles and found that the prediction and the actual measurement of that distance matched the prediction to within the width of a few human hairs!

With this introduction to our basic understanding of electricity and magnetism, then, we can turn to the question of how we can use this understanding to build a new kind of society.

THE ELECTRIC MOTOR

PROBABLY NO DEVICE is so widely used, and so widely unrecognized, as the electric motor. Motors are understandably associated with vehicles, but electric motors appear in things as simple as electric toothbrushes, blenders, and automatic windows in our cars. Electric motors are everywhere in our lives, doing all sorts of jobs.

But what, exactly, are they, and how do they convert the energy in electricity into the energy of a blender blade rotation or of a moving car window?

▶ How Motors Work

IT SHOULD COME AS NO surprise that the answers to these sorts of questions come from Maxwell's equations, particularly the equation that tells us that if we run electric current through a wire, we can create an electromagnet, as already discussed. As long as the current is running, this electromagnet will behave like an ordinary permanent magnet, with north and south poles. One important point to note is that if we reverse the direction of the current, the poles of the electromagnet reverse as well, with what was the north pole becoming the south pole and vice versa. This fact will become important in the operation of the electric motor.

Now consider the following: A wire loop carrying an electric current is placed in a magnetic field produced by a large permanent magnet. The wire coil is actually an electromagnet with its north pole corresponding to the north pole of the permanent magnet, and south to south. In this configuration, the north pole of the electromagnet is repelled by the north pole of the permanent magnet and attracted by the south pole of the permanent magnet. Similar forces act on the south pole of the electromagnet. Because of these magnetic forces, the wire loop rotates in a clockwise direction.

The rotation will proceed until the north pole of the electromagnet is opposite the south pole of the permanent magnet, at which point the attraction between the poles will stop the rotation. To keep this from happening, the current in the electromagnet is reversed before the north pole of the electromagnet reaches the point where the rotation can be stopped. Following this reversal, what was once the north pole of the electromagnet is now the south pole and vice versa. The new poles of the electromagnet are repelled by the poles of the permanent magnet, and the rotation continues.

The reversal of the current in a motor is carried out by a device called a commutator. Basically, the commutator is a split metal ring attached to the spinning shaft of the motor. Carbon blocks on the commutator, called brushes, provide an electrical contact (via wires, usually) to the electromagnet. These contacts are arranged so that the current through the electromagnet reverses as the shaft rotates. By changing the direction of the current in the rotating loop, then, this simple device produces exactly the changes needed to keep the rotation going.

▶ Going Electric

THIS IS THE BASIC OPERATING principle of the electric motor—a rotating electromagnet converts the electrical energy in the incoming current into rotational energy in the shaft. Of the many kinds of electric motors, each is designed to carry out a specific task. Most of them have two basic improvements over the simple design we described. For example, instead of one wire producing the electromagnet, many wires are bundled together to carry out this task. This allows for a more powerful rotation. In addition, most motors have more than one electromagnet, typically arranged at angles to one another. This provides a smoother rotation. Beyond these two designs, however, are literally dozens of others, each created to carry out a special task. As we advance through the world of electricity, we'll encounter several of these tasks, and we'll begin to piece together how the simple innovations of Volta and Maxwell led to things as seemingly complex as eVTOLs and Teslas.

Electric vehicle recharging stations like this need to become more common if we are to build a sustainable infrastructure that can support more electricity demand.

CHAPTER

6

Charging Up ... Everything!

M **MAXWELL'S EQUATIONS** embodied the connections between electricity and magnetism—connections that grant us the ability to generate electricity and transport it where we want to use it: our homes, businesses, and anywhere else we want to plug in something. But transporting electricity requires more than the basic knowledge that Maxwell imparted. We have to solve three problems when transporting electricity: (1) We have to be able to produce significant amounts of electricity; (2) we have to find a material that will carry electricity over a long distance; and (3) we have to deliver the electricity at a safe (that is, low) voltage. Let's start by seeing how we can use our basic understanding of electricity and magnetism to produce large amounts of electrical current, then figure out how to transfer that current to wherever it is needed.

The LEDs of Times Square cast electric blue light across an overhead view of midtown Manhattan.

INSIDE THE GENERATOR

THE FIRST ELECTRIC GENERATOR, built by Michael Faraday in 1831, worked according to the basic law of electromagnetism, but it really wasn't of much commercial use. The basic idea was that a metal disk was spun between the poles of a large magnet. Because the rotation created a changing magnetic field at the surface of the disk, an electric current was produced in the metal. Faraday arranged these components so that this current flowed radially to the edge of the disk, where a metal contact moved the current to an outside circuit.

Generators have come a long way since 1831. Modern generators produce what's called alternating current (AC) and are responsible for most of the electricity generated in modern societies. Alternating current refers to electrical current that changes direction on a regular basis—first it flows one way, then the opposite way. To understand how generators produce AC, imagine a large horseshoe magnet, which has a constant magnetic field in the region between the north and south poles of the magnet. Suppose further that a loop of wire (copper would be a good choice) can be rotated in that magnetic field. From the point of view of someone standing on that loop, the magnetic field is changing so long as the loop is rotating. (The strength of the field doesn't change, but the direction does, and this is enough to trigger electromagnetic induction.) Thus, as long as the loop is rotating, a current will flow in the wire—first one way, as the field increases, and then the other, as the field decreases. This is the AC—the electricity—that will eventually be sent to our homes: a simple example of an electric generator.

▶ AC Versus DC

BECAUSE THE ALTERNATING current reverses directions in the rotating loop, the current that reaches our homes will do the same. This AC is distinct from current that flows in only one direction—so-called direct current (DC). Direct current comes from batteries and solar photovoltaic cells (see pages 61–62). In the United States and elsewhere in the Western Hemisphere especially, the electrical system is set up so that the AC reverses itself 60 times a second, while in Europe, Asia, and Australia, most electrical

A generator found in the home of an electrician showcases the simple components that come together to make something that can power anything in our modern world.

An employee assembles generator products—updated versions of the designs seen above—at a workshop in Fuzhou, Fujian Province, China.

systems use an AC that reverses itself 50 times a second. Those figures have nothing to do with electrical efficiency or power—they're merely the result of historical choices from those who set up the systems. That frustrating difference is why global travelers need to carry adapters.

▶ Where Should Our Electricity Come From?

SO LONG AS WE HAVE a source of energy that can keep the wire loop spinning in the magnetic field, we can use this system to produce electricity. And this brings us to one of the central problems facing modern societies: Where is this energy to come from? In the case of electrical generators, we can identify many possible sources:

• **Hydroelectric power:** This type of energy is provided by water falling from a height, usually from a lake created by a dam. The water hits blades attached to the wire loop and keeps them rotating.

• **Fossil fuels:** These fuels, mainly coal and natural gas, are burned to pro-

An electric version of Maserati's MC20 racing-inspired sports car moves through a quality control tunnel.

duce heat that boils water. The resulting steam strikes blades attached to the rotating loop. The majority of electricity produced in the world comes from this source.

- **Nuclear fission reactors:** These installations use the heat generated by nuclear reactions to produce steam, which is used to keep the loop spinning. Nuclear reactors produce about 19 percent of the electricity used in the United States, and nearly 10 percent of electricity worldwide.
- **Wind:** Turbines use the energy of moving air. Basically, turning rotor blades keep the loop spinning to produce current. This process is described more fully on pages 64–65.

Solar energy is conspicuously absent from this list of energy sources. Why? Because even though solar can certainly be used to generate electricity, the solar energy itself comes in two forms, and only one of these produces an alternating current.

The most common form of solar energy in use today is photovoltaic solar energy, which uses semiconductors to produce DC from incoming sunlight, without the need for a generator. On the other hand, solar thermal systems employ mirrors to focus the sun's energy on a fluid-filled chamber that heats the fluid for steam to run a generator and produce AC. Thus solar thermal systems may become an AC power source, but they have a long way to go before being productive enough for modern demands.

HOW TO TRANSPORT ELECTRICAL ENERGY

ONCE THE GENERATOR has done its job, the next task is getting the electricity to the place where it will be used—our homes, for example. Electrical generating plants may be hundreds of miles away from where the current will be used, so we have to understand how different materials can (or cannot) carry electrical current over large distances.

All materials are made from atoms, and all atoms consist of a positively charged nucleus and electrons in orbit. If the nucleus of an atom was a bowling ball in a bowling alley, then the electrons would be like a few grains of sand scattered around the city where that bowling alley is located. That means an atom is mostly empty space.

When two atoms come together to form a material capable of

carrying electrical current, the outermost electrons in each atom will "see" the outermost electrons in the other atoms. These outermost electrons are called valence electrons. The way the atoms link up and the way the resulting material carries (or doesn't carry) electrical current depends on how those valence electrons interact with one another, and how they ultimately move through the material. We will look at insulators (which do not conduct electricity), semiconductors (the basis of modern electronics), and conductors (which, well, conduct electricity).

▶ Insulators

THE SIMPLEST PROCESS that can happen between two atoms is that one atom can give up a valence electron and another atom can pick it up. In this so-called ionic bond, each of the original atoms acquires an electrical charge—positive for the atom that gave up the electron, negative for the one that acquired it. A common example of the ionic bond is in ordinary table salt, where a sodium atom gives up an electron and a chlorine atom acquires it to make the material that seasons our food. Other materials, like ceramics, are actually held together by this sort of bond. Their electrons won't be dislodged from their molecular structure if the material is subjected to an external voltage. The electrons can't move, and so these materials will not conduct electricity. They are what we call insulators.

Another type of bond forms when two atoms shuffle a pair of electrons—one from each atom—back and forth between them. Each atom is essentially donating a valence electron to the exchange, so formation of this so-called covalent bond uses up one valence electron from each atom, two electrons total. Covalent bonds are common in many organic materials and plastics, and as a general rule, these kinds of materials are insulators and will not conduct electricity. Semiconductors, though, are an important exception to this rule.

▶ Semiconductors

SEMICONDUCTORS ARE the basis of the entire information revolution in which we are living. They're inside our smartphones, computers, robots—anything with digital programming. The most familiar semiconducting material is silicon, which has four valence electrons. In a piece of solid silicon (think a grain of sand), each silicon atom gives up its four valence electrons to form covalent bonds with its neighbors. It might seem that silicon is an insulator, because its valence electrons are locked in, but in fact the bonds are so weak that the ordinary jostling of atoms in the silicon can shake some of those electrons loose. Thus, silicon is called a semiconductor, with about a thousandth of the capacity to carry current as a fully conducting material like copper.

THE INVENTION OF THE TRANSFORMER

Engineers in England and France created the first prototype transformers in the late 19th century, but American industrialist and entrepreneur George Westinghouse (1846–1914) became eager to develop a commercially useful transformer after seeing an early demonstration. He assigned the task to his best engineer, William Stanley (1858–1916). Stanley's device ultimately contained a central iron core made of silicon steel laminations, with the wire loops shaped like an *E* and *I* so the copper could be wound easily into place. By running wires through the elm trees of his hometown in 1886, he completed the device and demonstrated its usefulness by lighting up the main street of his hometown, Great Barrington, Massachusetts. Later, Stanley formed his own company, which eventually became part of General Electric.

William Stanley's 1885 prototype transformer

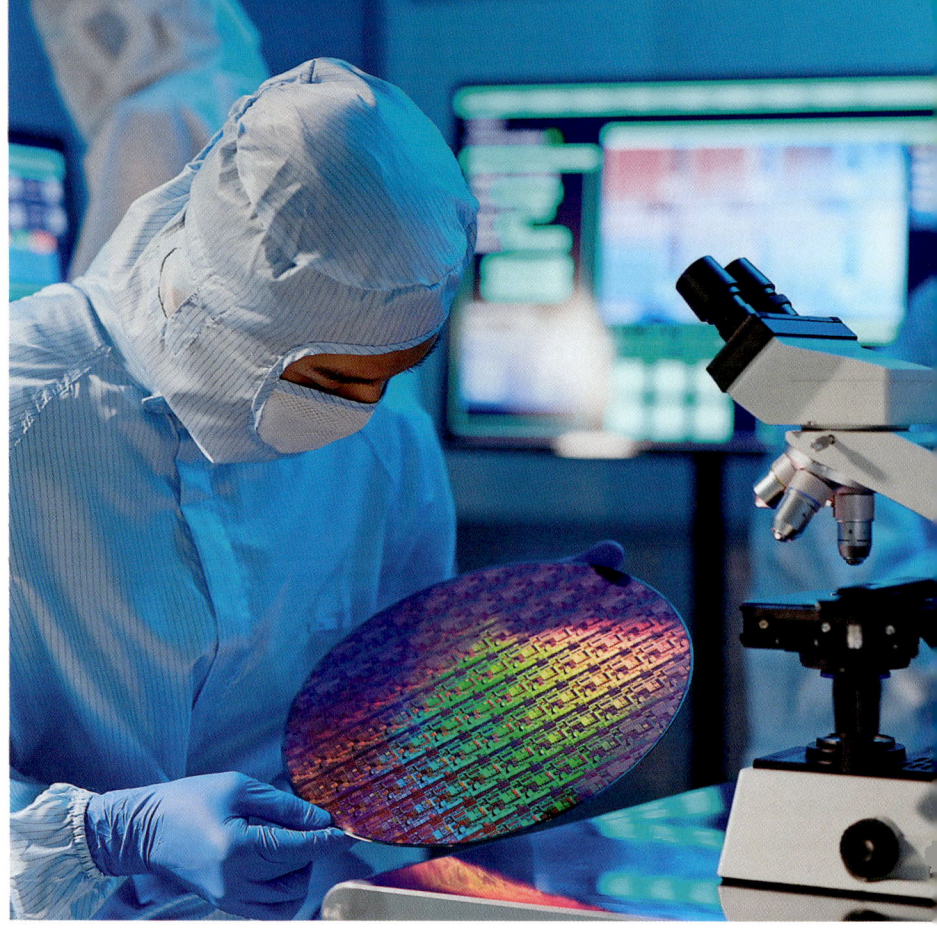

A technician examines a wafer of semiconductors, also called a slice or substrate, which can be used in solar panels or other electronics.

▶ Conductors

THIS BRINGS US TO METALS, the most common materials used to conduct electricity. Let's start by imagining a molten metal solidifying into a solid wire. When this happens, the metal atoms each give up one or more electrons, and the resulting ions lock together to create the solid metal. Meanwhile, the electrons that were given up form a kind of floating soup in the metal. Think of the metal ions forming a Tinkertoy-style solid matrix while the loose electrons (now called conduction electrons) form a kind of fluid around them. Insulators keep these electrons locked in tightly, but because conductors keep them free to move, they respond to a voltage.

Electrons carry a negative charge, and so when we submit a metal (or any other conductor) to a voltage difference, the electron fluid will start to move toward the positive voltage. This movement is the electric current.

Now, our conductor becomes highly useful for electricity distribution. With that current flow, it can carry electricity from one place to another.

ELECTRICAL RESISTANCE

AS THE ELECTRONS MOVE toward the positive voltage in a metal, they speed up and start colliding with the metal ions. In these collisions, the ions—locked in place by the processes that formed the metal—start to vibrate. In effect, the kinetic energy of the moving electrons has transferred to the ions, and we perceive the result of this energy transfer as heating of the wire. The process of energy loss—from the electrons to the metal ions—is referred to as resistance. High resistance means high energy loss, whereas low resistance means less energy loss.

If we're talking about transporting electrical current over long distances, we obviously want the current to flow through a material that doesn't lose much energy—a material that has low resistance, in other words. In fact, long-distance power lines usually use aluminum wire to take advantage of aluminum's low resistance—they can transfer a lot of electricity without wasting much energy. There are better conductors than aluminum—copper and gold, for example—but those materials are pretty expensive. (Financial efficiency, as we'll unfortunately observe, can be a higher motivator than material efficiency.)

But sometimes it's advantageous to have high resistance. Say it would be helpful to generate heat in a situation—like if we were making toast—then running electricity through a metal with high resistance would emit a lot of heat. That's why the coils in a toaster glow when it's on. A common material in toasters, space heaters, and hair dryers is a metal alloy called nickel-chromium, or Nichrome—roughly 80 percent nickel and 20 percent chromium, though it sometimes contains iron—that has a high resistance.

TRANSFORMERS (NOT THE ROBOT KIND)

WE'VE SOLVED HOW electricity gets to our homes, but now we confront another problem: How do we bring it into our homes and office buildings at a safe voltage? We've needed a high voltage to transfer it efficiently, but how do we use it without blowing our circuits?

Probably the best way to understand the problem of transmitting electricity over long distances is to think about water flowing through a pipe. In this analogy, the flowing water plays the role of electric current, the

A BOOST FOR BETTER POWER LINES

The same power lines have connected most of the world since 1908, but their design—an aluminum conductor around a steel core—is dated. Steel is heavy, and when it reaches 199°F (93°C), it stretches, making lines sag and limiting them in how much aluminum, or some other conductor, they can support. Less conductor means less peak power transmission: not good.

TS Conductor, a California-based public benefit corporation, is giving power lines an overhaul. Their bespoke core is made of a carbon fiber composite (the same stuff fighter jets are made of) that's one-fifth the weight of steel but can support 25 percent more conductor. Their design could facilitate at least double the energy transmission of a traditional power line, with less energy loss through resistance. And because this technology works within existing electrical infrastructure, it could be integrated tomorrow. "It is ready to deploy," says TS Conductor CEO Jason Huang. "It is ready to make an impact in energy transition. This tech can be the standard for the next 100 years."

The U.S. power grid comprises nearly 240,000 high-voltage transmission lines and 50 million transformers.

pressure pushing the water through the pipe plays the role of the voltage in the electrical system, and the friction between the water and the pipe plays the role of energy loss. The energy carried by the water depends both on the amount of water flowing and the speed with which the water flows: The higher the speed and flow rate, the more energy is carried, but also the more energy is lost to friction. As with the electrical system, the problem is maximizing the energy flow while minimizing the frictional loss.

The way to do this for moving water is to send as little water as possible through the pipe (minimizing loss) at the highest possible pressure (maximizing the energy transfer). Analogous reasoning tells us that the way to transmit electrical energy over long distances is to send the minimum current at the highest possible voltage.

To change the voltage of the electricity once it has been generated, we can turn to a device known as a transformer. Transformers appear in our lives every day (and not because they're robots in disguise). Each and every phone charger, for example, has one. And all the containers on power lines that look like garbage cans? Those have transformers inside them.

In a simple transformer, one wire carries an input voltage to one side of a square iron doughnut, or core. As this current travels through the coiled wire, it creates a magnetic field that crosses the doughnut to a separate wire coiled around the doughnut's far side. The presence of the magnetic field creates another current there, one that will produce its own magnetic field while flowing out of the transformer into whatever needs a charge.

Because the system must have equal energy input and output, engineers can manipulate the wires to create voltage differences in the two currents. If the input side of the transformer had 10 loops of wire, and the output side just five, the current would have to flow twice as fast on the output side at half the voltage. A system like this, in which the current flowing out of the transformer is at a lower voltage than that of the current flowing into the transformer, is called a step-down transformer. For the inverse scenario, engineers use the term step-up transformer.

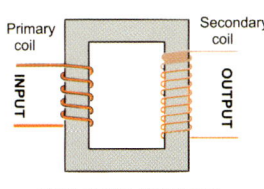

STEP-UP TRANSFORMER

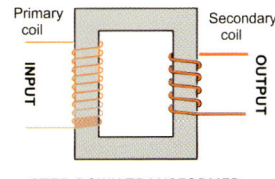

STEP-DOWN TRANSFORMER

▶ How Homes Receive Electricity

HERE'S HOW TRANSFORMERS make long-distance electrical transmission possible: At the generating plant, the voltage is stepped up to a very high currency—up to 765,000 volts in the United States. This high-voltage current is then sent out over long distances through the transmission lines that crisscross the country. When the high-voltage wires get to a city, the voltage is stepped down to a lower voltage (less than 10,000 volts) to be distributed locally. This takes place at distribution stations, which have become common in suburban neighborhoods. Finally, before the wires reach a given house, the voltage is stepped down again, to 240 volts, before it goes into a home's wiring system. These last step-down transformers are sometimes mounted on telephone poles on a residential street, but they can be buried too.

The current is brought into a house through the box that has circuit breakers or (in older buildings) fuses. Heavy equipment—clothes dryers, stoves, air conditioners—normally use the full 240 volts. Other circuits, like those that run our lights and TV, tap into half of the incoming voltage.

▶ Charging Our Phones

WE USE MODERN descendants of the simple transformer every time we charge our phones or computers. Here's why: A typical smartphone operates on direct current at 5 volts, while a typical laptop uses around 19 volts. Obviously, we can't just plug these devices into a wall socket—they would explode. Instead, the first thing that current coming from the wall encounters in our chargers is a sophisticated version of a step-down transformer. Immediately after the transformer adjusts the voltage, the current runs into another device in the charger. Called a rectifier, it converts the AC from a wall plug into the DC a smartphone requires. Then our devices can receive the charge.

We're already at a place where our most ubiquitous innovations run on powerful, efficient electricity. These simple building blocks—generators, conductors, and transformers—have essentially charged up the world! So where we'll see the most future innovation is not only in *what* we'll charge, but *how* we'll charge it. Electric cars and planes need more sophisticated versions of these building blocks than phone chargers or electric stoves do. In the next chapter, we'll look ahead to the future of giving things a charge.

The Mandarins electricity station at Bonningues-lès-Calais, France, sends converted electric current across the English Channel.

7

Lithium Batteries and the EV Takeover

ONE OF THE MOST significant steps in taking electricity everywhere is finding a place to put all the energy once at its destination. That, in essence, is what batteries do, and that's why they're so important to a future where our cars, trucks, buses, and trains are electric. Someday, most of us will be driving a car electrified by batteries under our seats, sharing the road with other battery-powered vehicles, perhaps cruising beneath battery-powered planes. Even our lives off the road will be impacted by batteries. When renewable energy sources like the wind and sun aren't active,

These lithium mine evaporation ponds in Silver Peak, Nevada, produce one percent of the world's total lithium supply.

batteries can have stored energy ready to keep things humming. As American electricity demands grow (some projections estimate we could add the usage equivalent of another California to the grid by 2028!), it'll become crucial to rely on battery technology so we don't have to return to fossil fuels for our energy needs.

| BATTERY BASICS |

ANYTIME WE USE AN electrical device that isn't plugged into a wall socket, we are likely using current supplied by a battery. In essence, a battery is a device that uses stored chemical energy to produce electrical current. Flashlights, cell phones, handheld gaming devices—even the electric aircraft Midnight that we met earlier in this section—depend on the operation of batteries.

In general, all batteries operate in pretty much the same way. The battery consists of two different kinds of materials, usually referred to as the cathode and anode, separated by a third type of material called an electrolyte. A chemical reaction removes electrons from atoms on one side of the battery (the anode), and these electrons flow through an external wire to the other side of the battery (the cathode), where another chemical reaction reverts them into atoms. We use that external flow of electrons to power our devices.

There are many different kinds of batteries, each relying on unique materials and reactions depending on its use. We'll describe a few categories here.

▶ Nonrechargeable Batteries

THE MOST COMMON TYPE OF disposable battery is called an alkaline battery. These consist of an outer cylinder made of zinc and a central spike made of manganese dioxide, with the two separated by an electrolyte. In the most common design, the electrolyte is made of ordinary lye (potassium hydroxide, KOH).

When a wire is connected between the zinc and manganese dioxide parts of the battery, the zinc reacts with material in the electrolyte to release two electrons, which travel through the wire: This is the current supplied by the battery. In this process, the zinc is converted to zinc oxide (ZnO) plus water. On the manganese side of the battery, the two electrons plus manganese form a molecule that has two manganese atoms and three oxygen atoms. This series of reactions continues until all of the manganese is used up, at which point the flow of electrons stops and the battery has to be replaced. This category of battery also includes the so-called button cells that power wristwatches, some types of hearing aids, and car fobs. Though zinc and manganese are the most common components, many different types of materials make up batteries, depending on the intended use case.

▶ Rechargeable Batteries

BATTERIES INTENDED TO be used sporadically over long periods of time are designed to be recharged, a process typically done by reversing the chemical reactions that produce electric current. Today, rechargeable batteries most often take the form of the lead-acid battery that starts a car and the lithium-ion battery that powers most smartphones and other devices.

Every car with an internal combustion engine needs a powerful kick to start the motor. Today, a rechargeable lead-acid battery provides that kick. The battery is built from two plates—one of pure lead, the other of a material made from lead atoms combined with two atoms of oxygen— immersed in an electrolyte of sulfuric acid. The chemical reactions begin with the lead giving up two electrons (these are what form the external electric current) and transforming into lead sulfate. (Lead sulfate is the crumbly white material sometimes seen around the terminals of old batteries.) At the other end of the battery, the electrons initiate a reaction that results in the plate being converted into lead sulfate and the sulfuric acid being converted into water. As the battery discharges, it turns into a system with two plates of lead sulfate immersed in water—a configuration from which no energy can be derived.

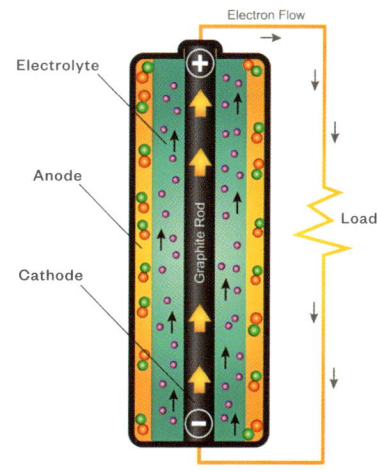

DRY CELL BATTERY

Electron Flow

Electrolyte

Anode

Cathode

Graphite Rod

Load

In a lead-acid battery, the chemical reactions we've just described can be run in reverse to return the battery to its original configuration. This is normally done by an electrical generator in a car's engine, which recharges the battery while the car is running. Thus, each time the car starts, it is—in principle—running on a fully charged battery.

▶ Lithium-Ion Batteries

LITHIUM-ION BATTERIES have become the workhorses of the information revolution. They power our phones, tablets, laptops, and most electric vehicles (EVs). They're also playing a growing role in renewable energy storage, where they stockpile electricity generated by intermittent wind and solar energy systems.

Lithium is one of the lightest elements in the periodic table, with a nucleus holding just three protons and four neutrons. Because of its atomic structure, lithium has a very easy time giving up one of its three electrons. These two properties, lightness and an easy supply of electrons, make it an ideal element to use in a battery.

The structure of a lithium-ion battery includes an anode—often made of carbon sheets—that provides a place where lithium atoms can be stored, kind of like an atomic parking garage. If a wire is connected between this anode and the battery's cathode, two things happen: The lithium atoms in the anode each give up one electron, which flows through the wire and is picked up by the cathode. At the same time, the lithium atoms minus an electron (making them lithium ions) move toward the cathode to balance the electric charge. The electrolyte between the anode and cathode allows the lithium ions, but not the electrons, to pass through to the other side of the battery. When all of the lithium atoms have moved through the battery and the cathode has absorbed the electrons, the battery is dead. But running

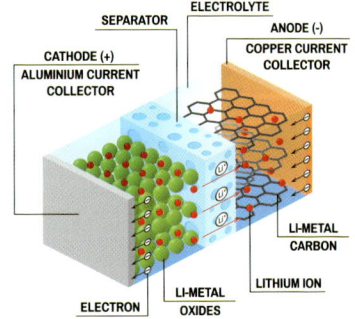

LITHIUM-ION BATTERY CHARGE

ELECTROLYTE
SEPARATOR
CATHODE (+)
ALUMINIUM CURRENT COLLECTOR
ANODE (-)
COPPER CURRENT COLLECTOR
LI-METAL CARBON
LITHIUM ION
LI-METAL OXIDES
ELECTRON

these reactions in reverse, returning lithium atoms to their parking garage, recharges the battery.

As would be expected for a device with so many uses, a number of bells and whistles are normally added to lithium-ion batteries. For example, to keep the anode and cathode structures from touching each other (which can cause the battery to explode), a semipermeable sheet inside the electrolyte, called a separator, is placed between them. This sheet

INTRODUCING SOLID-STATE BATTERIES

A As EVs occupy an increasing share of the vehicle market, automakers are racing to make a battery that can rival that of gas-powered cars: an all-solid-state battery (ASSB) with more energy density and quicker charge times than the lithium-ion battery in most EVs today.

Lithium-ion batteries contain a liquid electrolyte, but a solid-state battery uses a solid electrolyte, and the higher density of molecules in that solid, compared to that in a liquid, allows the battery to contain more energy. The exact material of the electrolyte is a trade secret for each company—this industry's version of the Coca-Cola recipe—but common materials include crystalline ceramics and glass-ceramics. QuantumScape, a leading ASSB developer partially owned by Volkswagen, boasts their tested battery achieves an energy density of 1,000 watt-hours per liter compared with 600 for leading lithium-ion batteries. The result for drivers? Four hundred miles (640 km) on a single charge compared to just 235 miles (380 km) on a lithium-ion battery—and the ability to recharge in just 15 minutes. The tech is enticing many automakers: Toyota and Nissan also have their own in-house ASSB development teams.

This rendering of a solid-state battery depicts the electrolyte in yellow between the solid electrodes.

In 1882, this German electric car constructed by Siemens & Halske would have been one of the first of its kind.

The first electric cars, like this electric cab introduced to London in 1897, weighed two tons and had a range of 30 miles (48 km) before they needed recharging.

also allows lithium ions, but not electrons, to pass through. Metal plates called collectors are usually placed near the anode and cathode as well. Finally, the layers of material that go into making a lithium-ion battery can be folded over to allow it to fit in a small space, and they can also be wrapped into cylinders to fit into the battery pack of EVs.

▶ Early Electric Vehicles

MOST PEOPLE ARE SURPRISED to learn that the first automobiles to travel our roads were electric vehicles, not gas burners. In fact, in 1900, more EVs were on American roads than gas-powered ones. It wasn't until Henry Ford began mass-producing the Model T in 1908 that the internal combustion engine became the leading mode of power for transportation in the 20th century.

The history of electric cars begins all the way back to the 1830s. Tinkerers in Holland, Britain, Hungary, and the United States put together chunky, clunky battery-powered vehicles that we would recognize as the ancestors of the modern automobile, but they were pretty primitive. This was before the invention of the rechargeable lead acid battery, so the electric cars of the era ran on disposable batteries. As you can imagine, they didn't travel very far. Their top speed was about 10 miles an hour (16 km/h).

▶ A Rechargeable Car Arrives

SCOTTISH IMMIGRANT William Morrison (ca 1850–1927) built the first commercial electric car in the United States in Des Moines, Iowa, in 1887. He was able to use the rechargeable lead-acid battery, which had been invented in 1859, and the newly developed electric motor to produce a vehicle that looked for all the world like a horse-drawn surrey—in fact, the passenger compartment of his car was built by the Des Moines Buggy Company. With front-wheel drive, a four-horsepower engine, and a reported top speed of 14 miles an hour (23 km/h), it had 24 battery cells that allowed a range of 50 miles (80 km) before recharging. Morrison's self-propelled carriage was a sensation at the 1893 Chicago World's Fair.

The last quarter of the 19th century brought a lot of progress in the electric car field. This period presented a four-way race for dominance in the field of personal transportation, with the contenders being EVs, steam-powered cars, cars with internal combustion engines, and horses.

In this competition, the EVs did quite well. One vehicle, called the Electrobat, even won a series of sprint races against cars powered by internal combustion engines in 1896. In the early 1900s, a fleet of 600 electric taxi cabs operated in New York City. Thomas Edison and Henry Ford even collaborated on an electric car for mass use.

▶ The End (and Rebirth) of the EV

THIS PROGRESS ENDED when Henry Ford began producing the Model T in 1908. In those days, it was easier to obtain gasoline than a supply of steady electric current, so Ford opted for an internal combustion engine in his car. A long period followed when the term "electric vehicle" conjured up pictures of things like golf carts, not roadworthy automobiles.

This started to shift in the 1970s, when the Arab oil embargo brought attention to the disadvantages of dependence on gasoline. When Tesla Inc. introduced their first all-electric vehicle in 2008, it heralded an approaching surge in EV sales worldwide. A negligible number of such vehicles were on the road in 2010, but as many as 26 million were by 2022. Optimistic projections assert that by 2030, the world fleet of EVs could exceed 350 million cars.

HOW AN ELECTRIC CAR WORKS

ELECTRIC CARS ARE powered by batteries, with the current from the batteries fed into a device called an inverter, which converts the current to the AC needed to run the motor. The motor used in electric cars is called an induction motor, invented more than a century ago by the polymath Nikola Tesla (1856–1943). This motor allows the operator to adjust the input current at will to change the current's frequency—essentially, the ability to give something "more juice" as needed. In a modern Tesla (named for that same inventor), each automobile has a battery pack containing more than 7,000 lithium-ion battery cells under the mainframe. These batteries are discharged when the car is in use and can be recharged by external electrical currents, a point to which we will return in a moment. As we saw earlier, lithium ions deliver direct current (DC).

There are several points to note about the induction motor. For one thing, it doesn't need a starter like the one in internal combustion motors. As soon as the current starts flowing through the external wires, the motor is on. For another, the induction motor operates at roughly the same efficiency no matter how fast it is turning. This means that electric cars don't need to shift gears in a complex transmission to accelerate.

UNDERSTANDING THE INDUCTION MOTOR

Electric cars are powered by a device called an induction motor. The motor has two key parts: a stationary structure around the outside, known as a stator, and a rotating shaft in the middle, known as a rotor. Current is sent through the windings of wire in the stator. This produces an electromagnet, and adjusting the input current can make this magnet rotate. Because the rotor is in a changing magnetic field, an electric current is induced in it, in turn creating another electromagnetic current.

The currents are arranged so that the poles in the rotor electromagnets chase the poles in the stator electromagnet. This causes the rotor to spin, and this spinning is what eventually drives the car. The frequency of the AC produced by the inverter controls the speed of the rotation: The higher the frequency, the faster the rotation—and the faster the car will go.

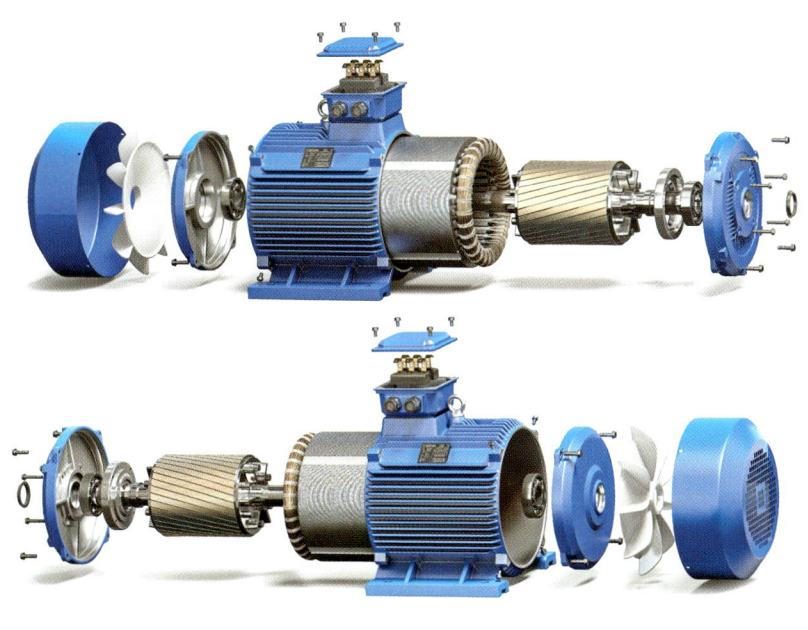

An exploded view of an induction motor, shown from two angles

A driver charges his electric
vehicle at a charging station
in California.

▶ When an EV Battery Dies

ELECTRIC CAR BATTERY packs need to be recharged by running electric current backward through the drivetrain. The distance a car can travel between charges is called its range. One of the major obstacles to the widespread adoption of electric cars is something known as range anxiety—essentially the fear that a car's battery will run out before it reaches a charging station. Improved battery technology may largely eliminate this problem—the Tesla Model S, for example, has an official range of more than 400 miles (640 km), about the same as a full gas tank in a compact car with a traditional engine.

Though filling up at a gas station takes only a few minutes, recharging a battery pack can take a long time. Plugging an electric car into an ordinary household 110- to 120-volt current overnight, for example, may add as little as 30 to 50 miles (48 to 80 km) to the car's range. The only way around this problem is to have an electrician install a special 220- to 240-volt charging system in the garage—an added expense associated with EVs.

Quick-charge stations are starting to appear in major cities, of course. These are estimated to proliferate, with (for example) as many as 12 million public stations and 28 million home stations projected in the United States by 2030, but at the moment, these stations tend to be few and far between. Anyone planning a long trip with an EV has to plan an itinerary built around these stations, and once they find them, recharging the battery pack may take 20 to 30 minutes—time that long-distance drivers hate to lose.

▶ Recharge by Braking

ONE IMPORTANT FEATURE of electric cars related to range issues is that they incorporate a phenomenon known as regenerative braking. Think about it this way: When a car is running, a vehicle weighing a ton or more is moving at some speed. A moving car, in other words, possesses a lot of kinetic energy. When the driver brings the car to a stop, that kinetic energy has to go somewhere. In a conventional automobile, most of it goes into heating the brakes and is eventually lost to the surrounding air. In an electric car, on the other hand, the energy in the turning wheels can be used to run the inverter and motor backward, in effect creating a generator that can be used to charge the battery pack. An estimated 60 to 70 percent of the kinetic energy of a moving electric car can be recovered in this way.

There is little doubt that technical problems like recharging time will eventually be solved, and that charging stations will someday be as common as gas stations are today. We may even find ways around the need for scarce substances like lithium to build car batteries. Assuming all of that, we will have to face a more fundamental problem with making EVs ubiquitous: Where will the energy come from to recharge all those cars? If it comes mainly from burning fossil fuels, as is the case today, the transition to EVs will have little effect on improving sustainability. The real transition will have to be coupled with the massive introduction of renewable energy sources, and a deep look at our electrical grid to see how it can support those systems. It sounds like an intimidating, daunting task—one that could take decades of stubborn persistence to achieve—but we're lucky that some of the best and brightest engineers in the world are already on the case. That's what we're exploring in the next chapter: building a smart grid that can keep up with all our phones, cars, and planes.

Stations like this one in Stockholm, Sweden, where Teslas wait at a charging station, will place new demands on electric grids and infrastructure.

CHAPTER

8

The **Future** of the **Power Grid**

D **DRIVING AN ELECTRIC CAR** doesn't feel much different from driving one that uses gasoline, and an electrically heated house can be just as warm as one that burns natural gas. Companies working on electric planes find it important to tout how similar the passenger experience will be on them compared to planes people are now used to. The main difference in electric vehicles is their silence compared to combustion engine vehicles. In fact, EV manufacturers are having to create artificial sound in their products to help users assimilate.

Transformers like these will need to become smarter to sustain increasing demand as humans use more energy from the grid.

More than a change in our lifestyles, our electric future will require a fundamental change in how energy is supplied. A glimpse at our all-electric future, therefore, requires that we look at how all our electricity is going to be generated and moved around.

UPDATING THE GRID

THERE IS A REASON that the world's largest solar electric installation is located in the Sahara desert and the world's largest wind farm is in the middle of the North Sea (see page 66). These vast, undeveloped landscapes are the kinds of places where generating electricity is easiest, but they tend to be far away from where the most electricity is used. So as our electricity demands grow, we will need to pay more attention to how electricity is delivered than we have in the past. Our power grid is going to be vitally important.

The grid that operates now is not the result of serious planning or design. It is, in fact, a relic of the 19th century. It was built piecemeal over the years, without much thought given to its overall efficiency, and certainly without recognition of its overwhelming importance to future centuries. It wasn't so much designed according to general principles as it was patched together as needed. This means that if we are really serious about building an all-electric future, we are going to have to think about the many improvements that will have to be made to the grid we have inherited. As often happens in situations like this, the best thing to do would be to tear down the grid completely and replace it with a well-designed and well-thought-out substitute. Because this is impossible, we'll have to do the next best thing—rebuild the grid we have.

▶ The Battle to Shape the Modern Grid

IN THE LAST DECADES of the 19th century, technologies like Thomas Edison's light bulb made it possible to use electricity in individual homes and businesses. Up to this point, tapping into electricity for personal use required a personal generator. Edison (1847–1931) decided to tackle this situation by building central generating plants and installing power lines to distribute the electricity. The problem was that his generators produced direct current. Because transformers don't work on DC, he was only able to get his current to installations a mile or so away from the generator. This meant that a city powered by Edison would require coal-burning generators every few miles—clearly not an inviting scenario.

At the same time, Nikola Tesla, backed by George Westinghouse, proposed something like our modern system, where alternating current (AC) generators could be located far from the places where electricity was to be used and delivered efficiently using a series of step-up and step-down transformers.

WHAT DID WE DO BEFORE THE GRID?

Before something like the grid existed, people who wanted electricity in their establishment essentially had to build a way to generate and distribute it themselves—a very exclusive expertise in those days. In Philadelphia, for example, before the invention of the light bulb, department store tycoon John Wanamaker (1838-1922) decided to install carbon arc lighting in his store to attract nighttime shoppers. He bought six generators and 28 carbon arc lamps and, in effect, set up an electrical generating system in the basement. The endeavor paid off for Wanamaker. On Christmas Day 1878, he threw the switch, lit up his store, and evening business went through the roof. (Wanamaker became quite the inventor. In 1889, he installed electric elevators in his store, and he later invented the price tag to eliminate bargaining for goods.)

Technicians discuss a studio arc spotlight in 1962.

A lot of money was riding on the choice of delivery systems, and the debate got pretty nasty. Supporters of Edison's system, in a smear campaign against Tesla, pointed out that AC was used to run electric chairs. They even electrocuted animals as part of their public lectures. In the end, the overwhelming efficiency and convenience of AC generation won out and gave us our present system.

But distribution via DC is making something of a comeback because of various technical advances. For example, the availability of doped semiconductors (see sidebar on page 60) has allowed us to build devices that can switch alternating current to direct current (and vice versa) with very little loss of energy. Consequently, the two types of current can be interchanged repeatedly in any grid. Finally, the details of the way that conduction electrons move in metal wires make it a bit cheaper to build long-range transmission systems that run on high-voltage DC (usually referred to as HVDC). In fact, most of the extremely long-distance systems we describe later in this chapter (see page 151) are of this type.

▶ Making the Grid Smarter

IN A TRADITIONAL GRID, electricity is sent out over long power lines at high voltage. When the lines get to a city, the voltage is stepped down for distribution and stepped down again before it is brought into a home. For most users, the only way to monitor the use of electricity is the monthly bill they get from their utility provider. (In fact, until well into the latter half of the 20th century, a power company employee would come by every month to read the meter in each individual home.) This lack of communication limited the ability of engineers to respond to short-term problems in the grid. When people talk about a smart grid, they are describing how tools from the information revolution could monitor and govern the flow of electricity—and thus get around this lack of communication.

The smart grid is one of the few technological goals that has been defined by an act of Congress. In 2005, President George W. Bush signed into law an Energy Policy Act, confirming the federal government's commitment to energy conservation and efficiency. It included the goal of a smart grid that would (among other things) increase the use of digital information,

The city lights of La Paz, Bolivia, stretch across roughly 182 square miles (471 km²).

Dubai's ever growing commercial sector accounts for about 50 percent of the city's total electricity consumption.

be equipped with full cybersecurity, integrate technologies to lower peak demand, give consumers real-time information, and remove unnecessary barriers to adoption so anyone could plug into it. The key point of these characteristics is the incorporation of modern information technology into our rapidly aging electrical grid.

▶ The Peak Demand Problem

TO TAKE ONE EXAMPLE of a problem with the conventional grid, consider the challenge of peak demand. When demand on the electrical grid

▶ **SOLVING THE STORAGE PROBLEM** ◀

The best example of a future-leaning storage solution, one that could be a template for how we make our grid smarter and more resilient, lies tucked away in the tranquil scenery of Virginia's Allegheny Mountains. The Bath County Pumped Storage Station powers 750,000 homes and balances the electricity needs of residential and commercial properties across six U.S. states.

Here's how: When electricity demand in the region spikes, valves in the station allow water—up to 13.5 million gallons (51.1 million L) a minute—to run from the upper reservoir to the lower reservoir through special tunnels. This turns six turbine generators that produce the electrical current needed to meet the excess demand. When demand settles again, the station pumps water back into the upper reservoir to prepare for the next spike.

The Bath County Pumped Storage Station in Virginia is one of the largest facilities of its kind in the world. Its net generating capacity is 3,003 megawatts.

peaks beyond what the grid can handle, the spike often leads to rolling brownouts—controlled reductions in grid-provided power—as engineers struggle to keep the electricity flowing. The present solution to this problem is to have standby generators for when demand is high. Real-time information transfer—a system that could tell the grid when an office park is in low demand on a summer Friday or a holiday Monday, for example—could be used to funnel resources to where they're needed most.

In another everyday example, residents of a given community on a smart grid and power suppliers to that community would both have access to on-the-spot information about how a residence's given appliances are functioning. A central computer might, during times of peak demand, shut off a refrigerator for 15 minutes so that energy can go somewhere more critical. A resident probably wouldn't notice this (and it wouldn't spoil their food!), but it could make a significant dent in the peak demand multiplied across all the users in a given area.

Engineers are already building bits and pieces of the smart grid. Some neighborhoods may already have smart meters that give their residents minute-by-minute information about the cost of the electricity they are using. Smart house apps could even allow homeowners to balance their own energy use through the day and night or during vacations. Other, more technical changes are in the works, and politicians are beginning to address the red tape around the grid. The question of who will pay for grid improvements is an important one, and people are already asking it.

▶ Hardening the Smart Grid

EARTH'S SURFACE IS a hard place to build an electrical grid. Extreme weather events routinely disrupt electrical service. The process of dealing with these sorts of threats is known as hardening the grid.

Hardening a grid against extreme weather is a relatively simple (but expensive) operation. Power lines can be buried, and the wooden poles that carry power lines can be replaced by steel poles set into concrete foundations. Guy wires installed in areas of high wind can keep poles and power lines from falling. Even simple maintenance tasks like removing trees that might fall on power lines can be thought of as hardening the grid against weather.

▶ Defense Against Geomagnetic Storms

GEOMAGNETIC STORMS are the result of emission of clouds of charged particles from the sun in events called coronal mass ejections (CMEs). (CMEs shouldn't be confused with solar flares—bursts of electromagnetic radiation from the sun that may or may not accompany a CME.) When these particles slam into Earth's atmosphere, they can produce cascades of charged particles raining down. From the point of view of a power line, charged particles represent an electric current, and we know from Maxwell's equations that two things will happen in this situation. First, the current associated with the moving particles will create a changing magnetic field in the region of the power line, and second, that changing magnetic field will produce a current in the power line.

That induced current will then run along the power line to wherever the line goes. If it comes into a house (as it can), it is called a surge. A surge can fry your computer—this is why we normally plug electronic devices into surge protectors. More significantly, however, the surge can be carried into the transformer stations integral to the electrical distribution system. If those transformers are burnt out, power can be shut down over an entire city. Given the present state of our manufacturing system, it could easily take a year to replace a damaged transformer. Think about how a city would fare without electricity for a year!

▶ Close Shaves With CMEs

LIKE HARDENING against extreme weather, hardening against a CME is a straightforward but expensive process. Typically, we have an early warning—sometimes several days—of an approaching storm, thanks to space- and Earth-based observatories. One precaution is to supply transformer systems with devices that can take them offline when a storm approaches. Another is to surround sensitive equipment with a metal mesh cage (called a Faraday cage) that plays the same role as a lightning rod, diverting surges away from the equipment and into the ground.

CMEs are actually fairly common occurrences. In 1989, for example, one shut down the entire power grid of Québec. The great-granddaddy of geomagnetic storms hit Earth in 1859, before a power grid was in place. Called the Carrington event (after the English astronomer Richard Carrington, who described it), the storm caused massive disruptions

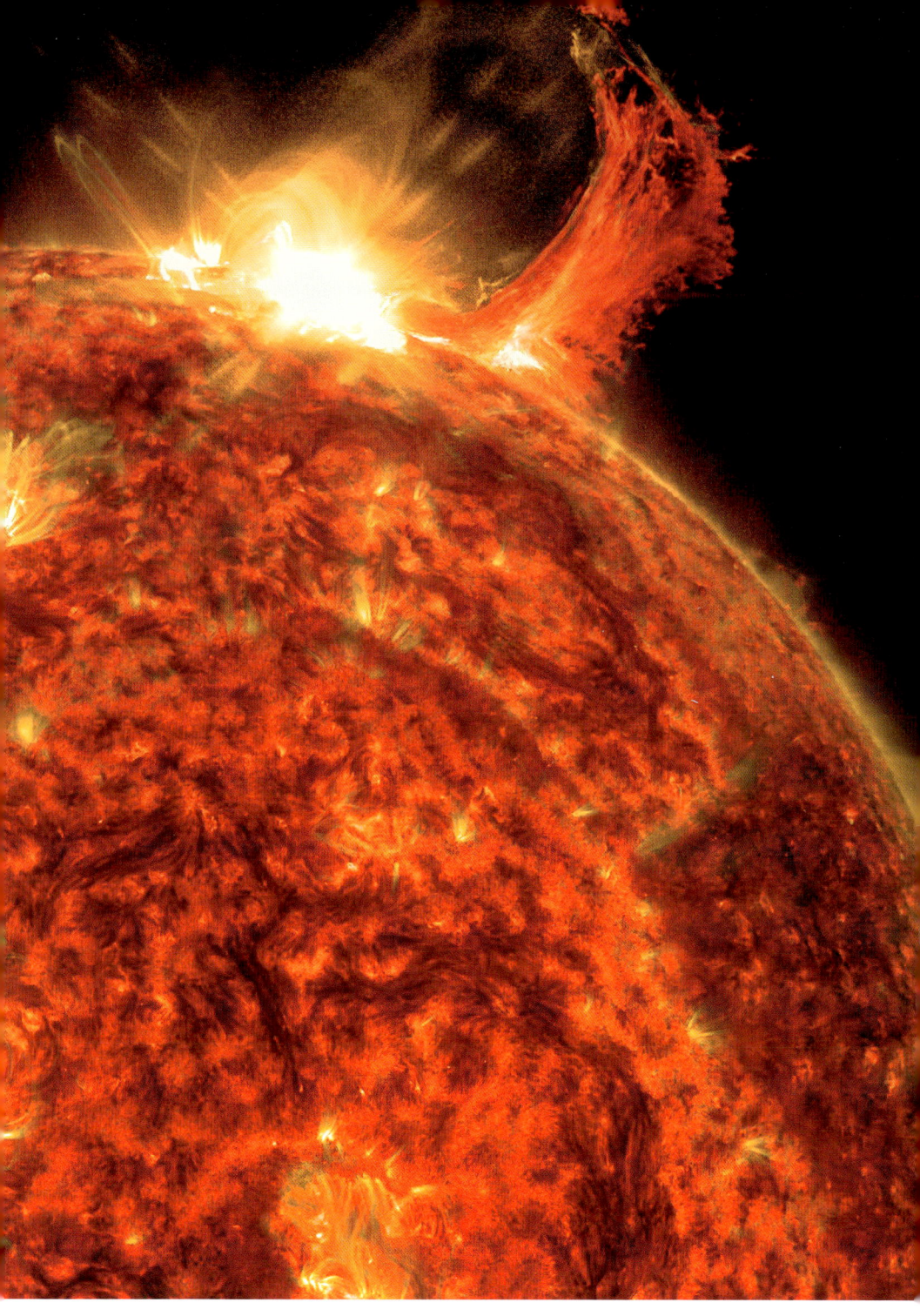

A solar flare unfurls from the surface of the sun. These flares can sometimes cause geomagnetic storms on Earth, which can lead to huge power outages.

to the nascent telegraph system. If something like the Carrington event hit Earth today, it would shut down power grids all over the planet—a scary prospect. In 2024, several CMEs contributed to the biggest geomagnetic storm on Earth in more than 20 years. While some power lines tripped and transformers overheated, we were lucky to avoid catastrophe—this time.

THE STORAGE PROBLEM

UNLIKE OTHER FORMS of energy, electricity must be used as soon as it is generated. The energy in gasoline can be stored until it's used, but electricity's energy can't be easily stored. Consequently, if we are going to use sporadic energy sources like solar and wind for our electricity, some of the electricity being generated must be stored for use when the sun goes down and the wind stops blowing.

We've already encountered such a storage system in the Noor solar thermal installation in Morocco (see page 61). Recall that in that system, some of the energy in sunlight is used to heat salts to a high temperature. At night, when there is no sunlight, that heat is used to boil water to create steam that powers the electrical generators. We should note, however, that although this sort of short-term storage works beautifully in the Sahara desert, it would have to be massively redesigned to generate enough electricity to carry a major city through several cloudy days.

▶ Batteries and the Grid

SYSTEMS CONSISTING of racks of lithium-ion batteries—enough to fill an 18-wheel truck—are already being put to use. The batteries are charged up when the sun is shining or the wind is blowing, and then are discharged when needed. Though a lithium-ion battery system like this will work, it is huge and very expensive. The cost of batteries that could keep a city functioning for an extended period of time could easily run into billions of dollars—far too much to be borne by individual users. Most likely, battery storage systems will function in the same way as emergency systems in modern hospitals, only occasionally and for short periods of time, with batteries providing minimal power until things return to normal.

▶ Pumped Storage and the Grid

PERHAPS THE MOST well-developed storage system is called pumped storage, a simple system that requires a particular type of topology. Imagine having two lakes in hilly country, with one lake at a higher altitude than the other. When demand is low during the day, some energy is used to pump water from the lower lake to the upper one. Then, as needed, water is allowed to fall back down to the lower lake, turning turbines that generate electricity. This simple system is known to work, but would obviously be difficult to create in a place with flat terrain.

Luckily, there are emerging battery designs that don't require special terrain. Iron-air batteries, for example, leverage the energy-storing potential of rust to provide the grid with power. These batteries use more readily available materials, aren't subject to thermal runaway (i.e., they won't explode), and can provide days of backup power. The start-up Form Energy finished construction on an iron-air battery factory in West Virginia in September 2024. They have projects in progress around the country and plans to build the world's largest battery in Maine to shore up the New England power grid.

▶ Can Wind and Solar Help the Grid?

THE STORAGE requirements for wind and solar energy—the two most important renewable energy sources—are quite different. Solar energy has one predictable variable—the sun rising and setting every day—and one unpredictable variable: cloudy weather. We need to generate at least enough excess energy during the day to power us through the night, but we also need to store enough extra energy to power us through days when clouds obscure the sun. So while future designs for these systems can prioritize building in desert regions (such as in Arizona and New Mexico in the United States), they should involve innovative long-distance transmission systems to route energy where it needs to go.

Though a great deal of long-term planning is possible for solar energy systems, the same is not true of wind energy. Although the United States has a huge wind resource, centered primarily in the High Plains, the availability of enough wind to top off energy storage systems cannot be assumed. Wind farms are usually located in places like the Columbia River Gorge along the Washington-Oregon border, where reliable winds

Rooftop solar panels power this building at SMA Solar Technology, a German solar energy equipment supplier and manufacturer.

These rechargeable battery blocks reside in a space the size of a school gymnasium in Schwerin, Germany, home of the first commercial battery park in Europe.

blow. These winds, however, are not as dependable as they seem. In 2009, for example, they stopped blowing in the Columbia River Gorge for three weeks! So far, no storage system that could get us through that kind of shutdown is in the planning stages. The only recourse would be a backup system run from a consistently available source, like fossil fuels or nuclear energy.

LONG-DISTANCE TRANSMISSIONS

ONE MORE TECHNOLOGY will have to be included in the grid of the future—one that can transport energy across thousands of miles. Think of a hot August day: All the air conditioners in New York are running, and solar energy is pouring into panels in Arizona. It would be great to send the energy harvested in Arizona to New York—but this task will require massive new infrastructure.

The longest power line in existence stretches 1,580 miles (2,540 km) across Brazil, from the hydroelectric plant in Belo Monte to Rio de Janeiro. For reference, this is about the same distance as from New York to San Antonio, and about two-thirds of the distance from New York to Phoenix. Thus, building the kind of transmission lines that would make the energy from sunlight falling on Arizona available to major U.S. East Coast cities would be a relatively small extension of existing technology— but a major regulatory and financial commitment. The thought of having to acquire the right-of-way to build such a power line is daunting, to say the least. The Grain Belt Express, for example—a transmission line meant to connect wind farms in Kansas to East Coast cities by 2028—will require permits from at least 20 state and federal authorities along the line's path. The line has to cross privately owned cropland, too, and understandably, not many farmers want a line carving through their harvest.

The good news about rebuilding the electrical grid is that the capability is well within reach: We have the technology to make our grid smart right now. And as things like batteries, electric vehicles, and pumped storage stations continue to grow, evolve, and prove their worth, the incentive to spend more resources on a better grid will increase in turn. Then, a lot of these bureaucratic difficulties will fade. We're already living in the electric future of our ancestors, with homes and cars and devices all plugged in. Soon, we could be living in our own electric future too—once we overhaul our infrastructure.

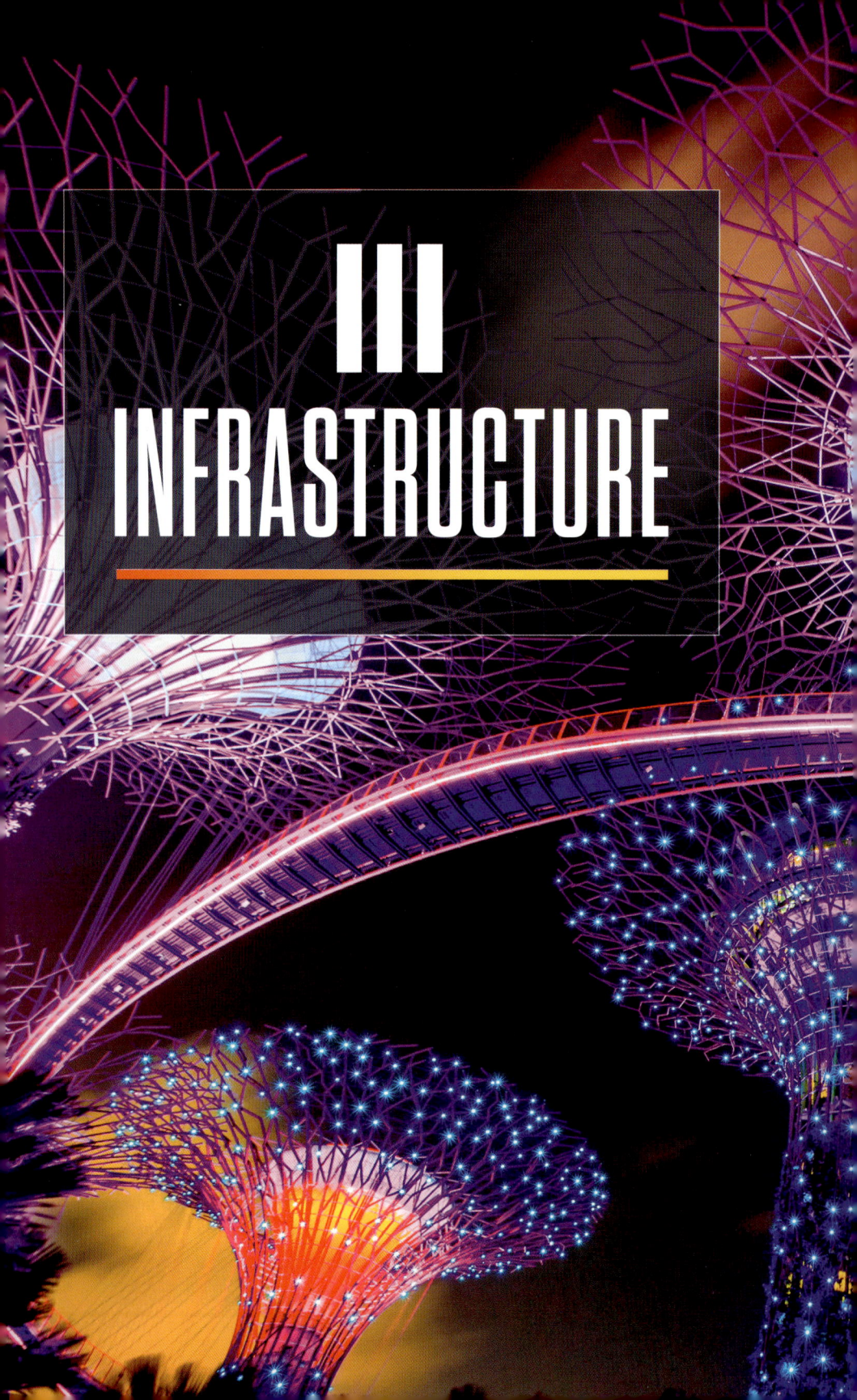

III

INFRASTRUCTURE

In Singapore, vertical gardens called Supertrees house more than 162,000 plants. Eleven of the Supertrees also harvest solar energy.

Touring an Ultramodern Wooden Skyscraper

I IN AN AMERICAN MIDSIZE city known most for breweries and Harley Davidsons stands the tallest timber-concrete hybrid building in the world. Nestled in Milwaukee's trendy North End neighborhood, between the city's serene namesake river and expansive Lake Michigan, is Ascent, a 284-foot-tall (87 m) luxury apartment building with a structural skeleton made mostly of wood. It's beautiful, evocative, and it likely represents the future of modern, sustainable infrastructure. When it comes to the materials we can use to expand and develop our cities, the old adage fits: "What's old is new again."

As our cities grow in land area and skyscraping heights, developers are beginning to push back (slightly) against the traditional building blocks of steel and concrete. Though these materials have undeniable construction strength and lasting durability, the sourcing and production of both can

Ascent, the tallest timber skyscraper in the world at the time of its construction in 2020–22, is 493,000 square feet (45,800 m²) and 25 stories tall. The building uses a six-level concrete podium structure for additional support.

have massive carbon footprints, and neither is renewable (yet—researchers are investigating "green steel" right now). Wood, meanwhile, is not only renewable, but strong and dependable. In fact, under the innovative architects and engineers building Ascent and other timber high-rises, it might be the best option for construction, period.

Ascent is the brainchild of Tim Gokhman. We meet on the sidewalk in front of his eco-friendly masterpiece. I gesture to the building's exterior. "It's beautiful," I say sincerely.

Gokhman smiles with satisfaction. "Thank you. It is, isn't it?"

Ascent is distinctly modern: The wood adds a note of comfort and sturdiness to the sleek transparency of glass. Called mass timber, the wood of Ascent is distinct from the heavy timber cut from whole trees you might imagine for a log cabin. Mass timber consists of small pieces of wood glued together to make a "massive" timber block, beam, or floor capable of supporting a tall, heavy structure. As Gokhman explained to me, mass timber holds several advantages for builders:

1. Environmental sustainability: Mass timber is sourced from renewable forests. You can't grow concrete or steel.

2. Cost savings: Timber is cheaper to transport than concrete or steel, and timber constructions move quickly thanks to the prefabricated components. The material itself is also thermally efficient—wood's natural insulation reduces heating and cooling costs in finished buildings.

3. Material strength: Mass timber buildings are lightweight and flexible, limiting seismic damage and enabling innovative building designs without the need for extensive foundations or structural support.

So basically, on paper, mass timber provides a clean, affordable, and reliable construction option with significant aesthetic benefits. But I want to see inside the walls. Gokhman leads me through Ascent's front door.

Nature greets us in the lobby via a lush live moss wall sprawling behind the doorman's desk. Silhouettes of fallen leaves dot the carpet, ceiling fixtures cast dappled light across the room, etched mirrors in the elevator evoke rain—no umbrella necessary—and, of course, exposed wood abounds, from

New Land Enterprises built Ascent in 22 months. Spokespeople for the building claim North American forests can regenerate the timber used for Ascent in about 23 minutes.

Engineers estimate Ascent will sequester about 7,940 U.S. tons (7,200 metric tons) of CO_2—the equivalent of taking 2,100 cars off the road for a year.

the building's massive timber columns to the cross-laminated ceilings in every apartment unit. The common area on the ground floor has minimalist, comfortable seating and a gas-burning stone fireplace. Piles of split wood, though only decorative, tower to the ceiling on either side of the hearth.

Naturally, the fireplace urges me to ask Gokhman the (ahem) burning question: Why are we building with wood if one match can bring down the whole thing? Gokhman isn't fazed. "Everyone asks that question," pulling me to a display of wooden pylons. These pieces were used to test the fire retardancy of Ascent's mass timber. Holding a 12-by-12-inch (30 × 30 cm) piece of wood (the largest in Ascent measures 4 feet by 4 feet [1.2 × 1.2 m]), Gokhman points out its charred border. He says the tester withstood three hours of torching at 1832°F (1000°C)! Even at that temperature, the burned outer layer extends less than an inch deep into the wood, leaving the timber at the center sound and untouched. Turns out, mass timber is so dense that—much like trees in a forest—only the outer layers burn, leaving the rest of the wood structurally unharmed.

Once I'm reassured, Gokhman invites me to sit in Ascent's work-from-home space, where we're canopied by a 25-foot-high (7.6 m) pine ceiling and bathed in natural light from the expansive windows. Downtown Milwaukee and Lake Michigan glitter below us as Gokhman tells me his story. He grew up in the former Soviet Union, present-day Ukraine, before fleeing to the United States with his family when he was just nine years old. After the family settled in Milwaukee, Tim's father, Boris Gokhman, used his civil engineering training to renovate homes in the area. The family pinched pennies until they could buy a small four-unit apartment building—their first real estate investment. Soon, they were able to leverage that property into another, then another. By the time they unveiled plans for Ascent in 2018, the Gokhmans had 31 buildings in their portfolio. Gokhman's eyes well with emotion when describing his family's journey. I ask what he's feeling. "Disbelief, pride," he says. "Somebody told me that if you believe something should be done, and no one else is doing it, then you're the person to do it. I guess that's our role."

Today, Ascent is a standout in forward-thinking, environmentally friendly luxury living. But like almost everything with hype, I imagine the downsides, or at least bad actors corrupting what should be a trailblazing model. Exactly how sustainable is mass timber? Is tree harvesting always eco-friendly? I wanted to investigate the philosophy behind Ascent, so I posed the question of Ascent's actual sustainability to Gokhman. I had my serious reporter face on, but his answer blew my mind:

"If you look at all the trees in all the forests in North America, it only takes 23 minutes of fiber growth to replace all of the wood we used to build Ascent."

Wait, what?! Only 23 minutes?! Since my arrival at the building that morning, six Ascents' worth of wood had grown in North American forests. That seemed impossible.

Gokhman's structural engineer worked with an organization called WoodWorks to arrive at that figure. The math gets a little complex, but it's backed with some peer-reviewed calculations of lumber-utilization percentages and growth rates. As a meteorologist, I can appreciate how highly specific formulas can coax computer models to pump out forecasts for nebulous things, like weather or wood growth. But still, 23 minutes demanded verification. I called Jeremy Whigham, professor of forestry at Alabama A&M University, to ask him about mass timber's sustainability and to fact-check the 23-minute claim. He chuckled, then said: "Well, it's probably hard to dispute, and certainly possible." When I checked the math myself, I found that 23 minutes was actually a conservative estimate. In the height of summer growing season, forests could produce an Ascent's worth of wood in less than 10 minutes! Sustainable? Check.

Ascent offers an alluring, and timely, promise amid infrastructure practices that depend on steel and concrete—reliable, but costly materials. For starters, steel production requires lots of heat—temperatures approaching 3000°F (1650°C)—so for every metric ton (1.1 U.S. tons) of steel produced, nearly two metric tons (2.2 U.S. tons) of carbon are emitted into the atmosphere. That equates to more than 7 percent of all greenhouse gas emissions in the world. Concrete has its own heat-intensive production process, but on top of that, the material itself absorbs tons of sunlight and doesn't absorb water, turning cities into "heat islands" and increasing the risk of floods.

Furthermore, most of the sand used to make concrete is pulled from crucial riverbeds—not good. The UN Environment Programme has reported that sand is the second most used resource on Earth, after water, with more than 55 billion U.S. tons (50 billion metric tons) extracted for construction *every year*. That's a lot of material— taken mostly from riverbeds and banks—carved away and repurposed for skylines. Changing a river's flow

Three species of wood make up Ascent—Douglas fir, black spruce, and Austrian spruce—all tested rigorously for fire safety.

Much of the natural wood used to build Ascent remains exposed, but it has all been tested to withstand direct flames for at least three hours.

can increase drought upstream and flooding risks downstream, and impact millions of people living near major waterways. Sand for other uses is mined from ocean floors and beaches. In 2019, the Yale School of the Environment says sand mining constitutes 85 percent of all mineral extraction in the world, and as a later section on deep-sea mining will reveal (see pages 188–191), the industry is far from clean. (Apparently sand mafias are a thing!)

For all these reasons, we need to diversify our construction practices—which may well involve a return to the age-old building material: wood.

Though the tree hugger in me sometimes wants us to live in yurts and tree houses, Ascent shows a new way forward. It's a hybrid building combining mass timber with steel and concrete in its foundation—on some level, that's just practical. And in the future, concrete and glass could become downright environmentally reasonable. Read on to learn how researchers in Canada are already producing and distributing a carbon-*negative* concrete and how a subdivision in Texas was built with 3D printers. There's hope for a future where we can build the things we need while stewarding the planet well. Ascent could be the model for how we meet those oft competing but ideally symbiotic goals. As Tim Gokhman says: "In construction, there is no one technology that solves everything. It's not like concrete and steel are going away. We just need to rebalance how and when we use them. So whether it's 3D printing or mass timber, [the future] is whatever gets us to carbon neutral and sustainable and beautiful."

Sounds pretty good to me—maybe I won't build my yurt just yet.

Ascent elevates and emphasizes its sustainable building materials by allowing lots of natural light into its 259 apartments.

CHAPTER

9

The **Pillars** of **Construction**

THE MAJORITY OF people in the world now live in cities, and it is unlikely that this growth of urban regions will slow down anytime soon. Our future, then, will involve a lot of choices about how cities will be built. In particular, it will involve choices about the kinds of materials we will use to build our future dwellings and places of business. Some of these materials will be familiar, but, like Ascent and other timber skyscrapers, they will be new versions of old songs. Other materials will be things that aren't common today—or even things we haven't built with before—but they'll appear as stronger, more sustainable, or more available alternatives to the things we use already.

Beijing National Stadium, or the Bird's Nest, is one of the largest steel constructions in the world.

BUILDING WITH CONCRETE

HUMANS HAVE USED concrete since Roman times, and we now use about 33 billion U.S. tons (30 billion metric tons) of the stuff every year—enough to build several thousand Hoover Dams. Our use is accelerating too: Since 2003, China (the world's largest user and producer of concrete) has poured more concrete every two years than the U.S. poured in the entire 20th century! Concrete will be in our future. It's in our highways, dams, sidewalks, bridges, and so much more.

▶ WHY DOES ROMAN CONCRETE LAST SO LONG? ◀

The ancient Romans were remarkable engineers. Roman concrete structures still dot the Mediterranean more than a millennium after they were built— some of their aqueducts even carry water into Rome today, and the dome of the Pantheon, built in 128 c.e., still stands.

It used to be thought that the special durability of Roman concrete was due to the ground-up volcanic rocks they used for their aggregates. Recently, however, researchers at Massachusetts Institute of Technology (MIT) have discovered millimeter-size white objects throughout the ancient concrete. Called lime clasts, these mineral bits perform a specialized function. If a crack opens in the concrete, water gets in and the clasts fracture. The calcium in the clasts then recrystallizes and fills the crack. Thus, the Roman cement, in essence, heals itself. Because modern concrete might last only 50 years by comparison, chemists are searching for ways to incorporate Roman ideas into modern concrete mixtures.

Roman concrete has helped the Colosseum stand for nearly two millennia.

Building the Glen Canyon Dam on the Colorado River required more than 400,000 buckets of concrete. Each bucket held 24 U.S. tons (22 metric tons) of concrete.

But what, exactly, is concrete?

Basically, concrete is a mixture of water, cement, and a material called an aggregate (sand, gravel, or crushed stone). The water and cement form a slurry that incorporates the aggregate—this is the concrete you see being poured. As the water is absorbed, the cement gradually fills the spaces between the aggregate particles. The end result is a hard, stonelike substance that's strong, versatile, and cheap to produce.

Cement is the key ingredient in making concrete, and since its development in the 19th century, the material of choice has been a substance called Portland cement. (The name comes from the cement's color—it's similar to that of stones found on Portland Island off the coast of Britain.) Portland cement is made by heating a mixture of limestone and clay minerals to a high temperature for a long time, typically in a large kiln. This step in the production of concrete raises environmental concerns, because the heating process is usually carried out using fossil fuels. Consequently, the production of the cement needed to make concrete adds carbon dioxide to the atmosphere. In fact, because of the widespread use of concrete in construction, as much as an estimated 8 percent of human carbon emissions in 2022 came from production of cement for concrete.

▶ Stronger, Brighter Concretes

IT IS EXTREMELY UNLIKELY that humans will give up using concrete—it's just too convenient. That doesn't mean, however, that it can't be improved. In view of the large amount of concrete in construction worldwide and the large carbon footprint that accompanies it, scientists are exploring a number of avenues to develop a concrete of the future.

Ultra-high-performance concrete (UHPC) was developed in the late 1980s by the U.S. Army Corps of Engineers, and brought to market by a French company soon after. This type of concrete uses a unique, very detailed recipe for cement that calls for very little water and the inclusion of fibers from steel, carbon, or several other materials, which are added to the cement as it is mixed.

Ordinary concrete usually shatters between 400 and 700 pounds per square inch (28–48 bars), while UHPC doesn't fail until the pressure reaches 1,400 psi (97 bars). In addition, UHPC handles weather much better than conventional concrete. It is expected to last up to and beyond

100 years as opposed to fewer than 50 years for conventional concrete, and that means less time, money, and resources spent on repairing a given structure—good for a future where we build smarter than we do today.

Scientists are also trying to produce something we can think of as a glow-in-the-dark concrete. To produce such a material, they mix in small amounts of titanium and sulfur compounds when the cement is being readied for use. When the concrete is poured, these inclusions absorb sunlight, then emit light after dark. Think of a sidewalk or a road built from this kind of concrete. It would light up after sunset, and you might need fewer streetlights. Cross-country highways might be safer for nighttime driving, and residential streets might be safer for vulnerable populations. At the moment, scientists are experimenting to find the best combination of additives to maximize the effectiveness of the technique.

▶ Concretes That Work for the Planet

GRAVEL AND SAND AREN'T the only aggregates that can be used to make concrete. The Romans, after all, prized the ground-up volcanic rocks from the Naples area. Today, sustainability efforts have initiated a search for materials left over from other processes that might be used to make new concrete. A current favorite is fly ash, fused impurities produced when coal is burned in electric power plants. It appears that concrete made with this material injects less carbon dioxide into the atmosphere than concrete made with Portland cement. In addition, concretes made with fly ash turn out to be highly durable.

Perhaps most exciting of all, modern engineers are taking an idea out of the ancient Roman playbook and experimenting with systems to heal the inevitable cracks that develop in concrete over time. Instead of using the Roman technique involving lime clasts (see sidebar on page 166), however, modern self-healing techniques incorporate dormant bacteria and calcium-based nutrients into the slurry when concrete is being formed. When (not if) a small crack opens up, allowing water to enter the concrete, the bacteria begin converting the nutrients into insoluble limestone (calcium carbonate), thus sealing the crack. Literally dozens of bacterium species can be used to do this job.

All of these advanced concretes carry immense promise, but at present, they come with expensive price points and fairly sophisticated assembly and mixing processes. This means that in the near future—and at least until these problems are solved—their use will most likely be confined to major construction projects.

| BUILDING WITH STEEL |

WHEN WE THINK OF tall buildings, we usually picture those with a steel-frame skeleton to support the weight of the structure, like the Willis Tower in Chicago, the tallest steel-frame structure in the world. It would be hard to imagine the skyscraper boom of the 20th century without the widespread availability of steel—even though, as pointed out previously, high-performance concrete is replacing it in current construction.

Steel is an alloy of iron and carbon. Iron on its own, like most pure metals, is not a particularly strong material, but creating an alloy addresses that problem. Iron must be brought to a liquid form at high temperatures to add other metals to it and create an alloy. Think of the metal as something like a bunch of marbles in a sea of molasses. When it cools and solidifies, the individual atoms give up electrons that form a kind of sticky liquid that holds the atoms together. The addition of a few carbon atoms to molten iron serves to pin down the iron atoms, creating a material that is much stronger than the iron by itself.

For most of human history, steel was a rare commodity. The reason was simple—most modern manufacturers introduce carbon into a piece of iron by melting the iron, pouring in the carbon, and letting the mixture cool and solidify—but you need a furnace capable of reaching temperatures above the melting point of iron, 2800°F (1538°C). Until the mid-19th century, this sort of temperature was simply beyond the capability of existing technology. But in 1856, the English industrialist Henry Bessemer (1813–1898), in search of ways of making better cannon barrels, developed a process that, by blowing air through a mixture containing iron and some impurities, was able to maintain temperatures well above what is needed to melt iron. Later on, engineers built on Bessemer's idea by blowing pure oxygen through a mixture of iron, scrap steel, and impurities like limestone to produce steel much more quickly in what is called the basic oxygen process. This process produces most of the steel we use today.

▶ Custom Steel Recipes

REALIZING WHAT A CRUCIAL change these inventions made is important. A single blow of a Bessemer converter, taking only about 15 minutes, could produce 10 to 25 U.S. tons (9–23 metric tons) of steel—more than what was available in the entire Roman Empire. In addition, once producing large quantities of molten iron was possible, people discovered that adding other materials to the mix produced steels with different qualities. Add chromium, for example, and you get a steel that won't corrode—stainless steel. Add tungsten and you get a steel that won't lose its edge when heated

A worker cleans a melting pot at the Salzgitter AG steelworks, one of Europe's largest steel producers, in Salzgitter, Germany.

and can be used as a cutting edge in industrial machinery. Add vanadium and you get a steel tough enough to be used in railroad tracks. Some steels today contain more of these additional atoms than atoms of iron, but the most common steels—the ones used in buildings—have only carbon as an additive. It doesn't take a whole lot of carbon to do the job of holding up a skyscraper—most construction steel is only about 0.15 to 0.3 percent

NORWAY: OUR LEADER TO A TIMBER FUTURE?

Ascent's most influential predecessor among timber skyscrapers might be Mjøstårnet in Brumunddal, Norway. Engineers completed this 280-foot (85 m) Scandinavian structure in 2019 using 16,000 spruce trees, and they set an enviable precedent. For every tree harvested from Norwegian forests to build Mjøstårnet, two trees were planted, says Rune Abrahamsen, former CEO of Moelven Limtre AS, which provided Mjøstårnet's timber elements. It takes 80 years for these new trees to grow, but Abrahamsen points out that a timber structure can last for centuries in the right conditions—in fact, several of Norway's churches have survived with their original wood since the 1100s. Will Mjøstårnet and Ascent see the 2900s?

If they do, they might be joined by a figurative forest of timber skyscrapers around the world. The United States housed 2,427 completed or in-progress mass timber projects as of March 2025, according to WoodWorks, the timber-construction consulting firm that contributed to Ascent. Timber's practical appeal is growing, while its green appeal persists. "We look at ourselves as advocates when we build a building like this," Abrahamsen says. "It's our hope we can influence others around the globe to do likewise."

Mjøstårnet, another tall timber building, on Norway's Lake Mjøsa

carbon, for example. Most of the construction steels used today are recycled, containing the necessary carbon already.

One modern development likely to be important in the future are the so-called high-strength low-alloy steels. In addition to carbon, these steels contain an alphabet soup of various elements: copper, nickel, niobium, nitrogen, vanadium, chromium, molybdenum, titanium, calcium, rare earth elements, or zirconium. These special steels are currently being tested in highway bridge construction, where their added strength and resistance to corrosion are obvious advantages.

BUILDING WITH WOOD

WOOD IS PROBABLY THE oldest building material that humans have used. It is easy to acquire, easy to shape, and easy to fasten into structures. Even today it is the most common material for buildings only a few stories high, and as we saw at Ascent, it might become more common in high-rises before too long.

The reason wood buildings have been of limited height in the past has to do with the structure of wood itself. Think of a piece of wood as being analogous to a handful of drinking straws. It is possible to balance a reasonably heavy weight on top of the handful of straws, but if you push sideways on the straws, the handful will collapse. This is because a sideways push will cause one straw to buckle and exert a sideways force on the next straw, which will also buckle, and so forth. In the language of engineers, we say that wood is strong against compression (the downward force) and weak against shear (the sideways force). Build a conventional wood structure too high, and the building could collapse the first time the wind exerts a shearing force.

▶ Wood's Fresh Growth

IN THE LAST FEW DECADES, however, a new process has been developed that has returned wood to the forefront of high-rise construction, and made buildings like Ascent possible—using a material we all know. Plywood is made from sheets of wood that have been shaved off a tree, more or less the way a roll of toilet paper unrolls. The wood in each sheet is strong in the direction that the grain runs, but weak in the direction perpendicular to the grain. This weakness is overcome by gluing sheets of wood together with alternating grain orientations, so that, no matter the direction of the applied force, a sheet will always be oriented in a way that opposes it. The final product, sometimes known as cross-laminated timber (CLT), has become a staple in large-scale timber constructions. Ascent uses CLT in its apartment units, ensuring that the building can withstand weather impacts upon its uppermost floors.

Many engineered wood components like CLT have been developed that make use of modern epoxies or other adhesive techniques to join pieces of wood into strong construction materials. Glulam, or glued laminated

The Metropol Parasol in Seville, Spain, is made entirely of wood. It towers 85.3 feet (26 m) and features a special walkway on top for visitors.

timber, uses moisture-resistant adhesives to join sheets of wood, whereas dowel-laminated timber (DLT) uses dowels to fit its layers together.

Ascent is not the only skyscraper of its kind. High-rise structures have been built from engineered wood in Europe, North America, and Australia. An important feature of these structures is that carbon removed from the atmosphere when the timber was growing stays in the structure, making it more green than CO_2-producing concrete.

| BUILDING WITH GLASS |

GLASS IS ONE OF THE oldest manufactured materials used by humans. It is a fairly simple substance, basically a collection of silicon and oxygen atoms. In principle, this means we could make glass by melting sand, because sand contains both silicon and oxygen, but the process would require a temperature of several thousand degrees Fahrenheit. Consequently, when glass first started to be made thousands of years ago, soda ash (sodium carbonate) was added to the sand to lower the melting point. Limestone was also added so that the final glass wouldn't dissolve in water. This combination of soda ash, limestone, and silica (sand) remains the formula for making ordinary glass today.

The modern method of producing glass, developed in the mid-20th century, is called the float glass process. Molten glass is poured over a container of molten metal (usually tin, as it has the optimal density). Gravity causes the layer of molten glass to become a sheet of constant thickness. The glass is then cooled and cut. The glass windows of your home or car were likely made this way.

Glass has been part of human technology for millennia. Today, it has assumed a more central role because of the availability of steel and concrete skeletons for modern buildings, as discussed. It is not at all unusual to see modern skyscrapers with exterior walls made entirely of glass, because the steel or concrete frame supports the weight of the building.

▶ How Glass Gets Smart

IN MODERN BUILDINGS, glass plays the role of the outer skin and therefore governs the way that the building interacts with the surrounding climate, weather, and air quality. In fact, a new type of glass called smart glass is being developed to allow this very old material to play a new role in regulating the energy flow in and out of buildings.

So-called smart glass is designed to control and adjust the way light passes through windows into a building. On a hot summer day, it's preferable to reflect light to keep the building cool, while on a cold day in January, it's preferable for the light to help warm the room. Smart glass can accomplish both of these tasks.

Two panes of glass (like two pieces of bread) make essentially a smart glass sandwich. The material between the glass panes is filled with long molecules like the ones moved around to create the crystal display in a smartphone. At rest, when no voltage is applied to the glass, the molecules will be randomly oriented to prevent light from coming through the window, making it opaque. But when voltage is applied to the system, the molecules line up and allow light to get through, making the window transparent. Thus, depending on how the voltage is applied to a window, you can control light coming into the room, thus having more control over the climate of the room itself. (The voltage has to be applied only once to set the transparency of the window—the molecules stay aligned until a new transparency is desired—so the window actually uses very little electricity.)

Many other systems are being developed for smart windows, but all of them work thanks to the meat of the sandwich, as in this example. They differ only in how the transparency of the material in the middle of the sandwich is controlled. For the moment, the bread in the smart glass system remains pretty much the same as glass we've been using since we first discovered the material.

Although smart glass operates on attainable, replicable technology and material, it can still be very expensive. A smart glass window, for example, can cost 10 times what an ordinary window does. Over a huge skyscraper, this represents a formidable obstacle to builders and designers. Most of these sorts of upgraded materials we've been discussing—from self-healing concrete to unique steel alloys—face significant cost barriers as well, but as we're about to see in the next chapter, emerging markets for rare materials that outfit these technologies could one day level the playing field. We just need to find the right techniques to safely and efficiently harvest them without harming the planet.

The glass of Harpa public concert hall in Reykjavík, Iceland, is designed to evoke the country's crystallized basalt columns.

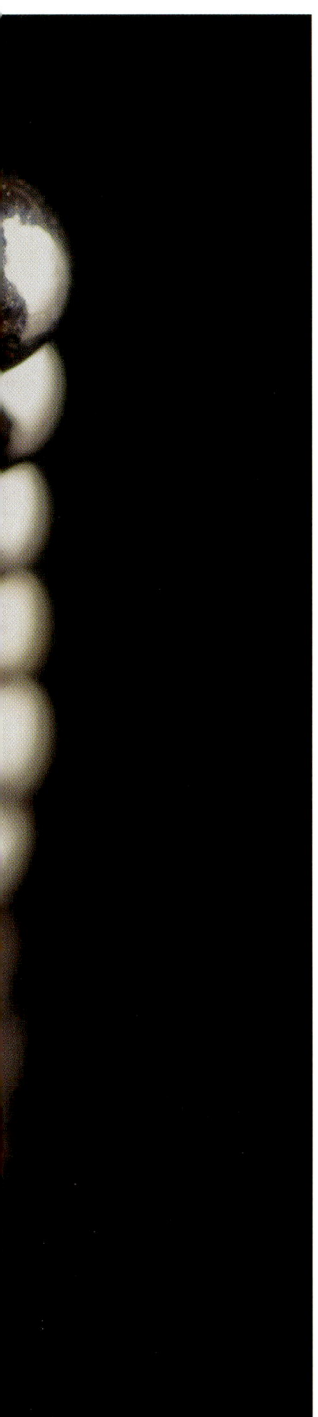

CHAPTER

10

Rare Metals and Special Materials

THE GREAT ADVENTURE we have embarked on to develop a future-ready infrastructure, with more sustainable concrete, timber, and glass, will require new materials to make the innovations work as they should. These materials are sometimes referred to as neo-commodities.

Particularly as our world grows more electric, minerals that nobody needed 20 years ago are suddenly on everyone's radar. Electric vehicles, for example, are going to require vast amounts of lithium and cobalt—metals mostly overlooked in transportation and construction—to reach

Neodymium—seen in this cube of neodymium magnets—is a rare earth metal used most commonly in electronic devices.

the level of ubiquity that certain sectors of the auto industry are hoping for. In this chapter, we'll look at a few of these minerals and explain why they're inspiring a new, 21st-century gold rush. We'll also look at the problems we're going to face in ensuring an adequate supply of these minerals in the future.

LITHIUM: THE NEW GOLD

LITHIUM-ION BATTERIES are the workhorses of the information revolution, and they will only become more important as our world becomes more digitized. Lithium-ion batteries power the consumer electronics devices we use every day, and are the main energy source for electric vehicles. In the far future, they could be central to the energy storage systems developed for renewable but sporadic energy sources, such as wind and solar. Right now, lithium-ion batteries rank alongside pumped storage hydropower and iron-air batteries as our best solutions to this technological problem. But that means demand for lithium-ion batteries will skyrocket. Is there enough lithium available to meet all of that demand?

The world's proven lithium reserves stand at 33 million U.S. tons (30 million metric tons)—well above the amount of lithium likely to be required in the near future. Some estimates assert the world will need 2.5 million U.S. tons (2.1 million metric tons) of lithium annually by 2030. Conventional lithium mining poses severe environmental problems, however, and refining lithium ores into the pure-grade substances required by batteries is its own messy problem.

▶ Lithium Mining Today

LITHIUM IS NOT A PARTICULARLY rare element. At the moment, it's mined in Australia, China, and the so-called Lithium Triangle, where Argentina, Bolivia, and Chile meet. When the mining industry realized how intense the demand for the mineral is going to be in the future, they invested in the discovery of new sources of lithium all around the world. But whether new mining operations can actually start producing lithium in time to supply our growing technological needs is a subject of debate.

Once a deposit of lithium is found, it can be mined in two ways. One

Evaporation pools at the Llipi lithium plant in Bolivia contain one of the world's largest concentrations of lithium.

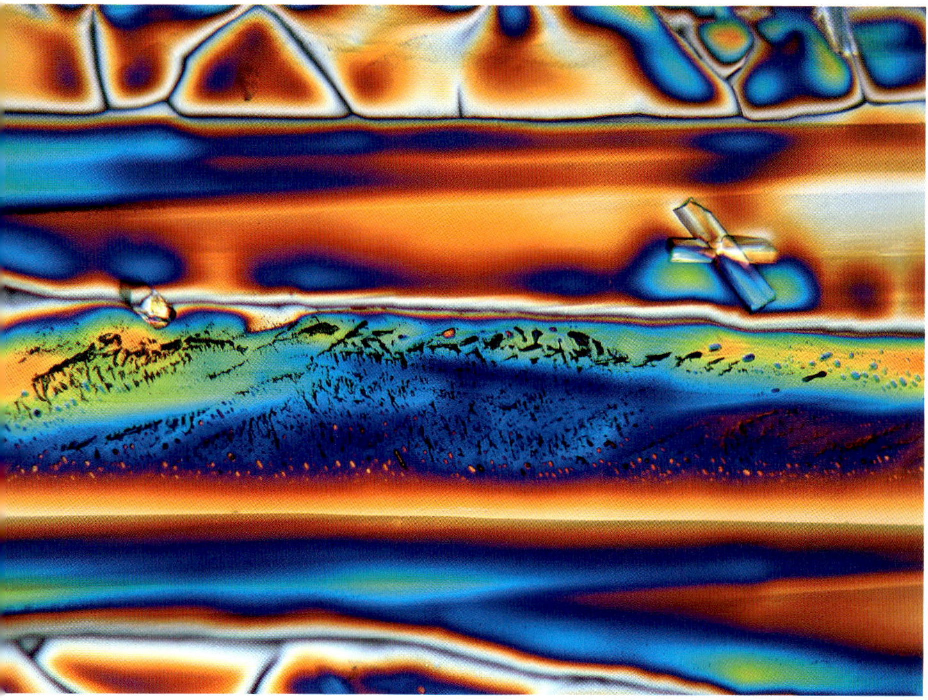

Magnesium sulfate, a common leaching agent used to help mine rare earth metals, as seen under a microscope

(predominantly used in Australia) is conventional hard-rock mining, in which chunks of lithium-bearing minerals are blasted out of tunnels and brought to the surface. The other technique, used in the Lithium Triangle, involves the use of brines. Here's how it works: Underground water leaches lithium-rich salts, and the water is then pumped to the surface and allowed to evaporate. The evaporation process leaves behind the lithium. At first glance, it would seem that recovering lithium from brines, a very efficient process, is an ideal way to recover underground lithium, because minimal excavation and minimal damage to Earth's surface are involved. The process, however, raises problems. The brine technique is currently used in regions where water is scarce—a fact that can lead to conflicts between miners and local populations who rely on the water for survival. Furthermore, the land area required for the evaporation process can be quite large: Brine ponds in the Lithium Triangle are often the size of Manhattan. In addition, the process is slow. It is not unusual for the evaporation process to take a couple of years.

▶ The Tricks of Refining Lithium

ONCE THE LITHIUM-RICH ORE is mined, another problem arises. The ore must be refined to produce lithium of a purity required for batteries. Ores obtained from conventional hard-rock mining are processed in the usual way: The ore is ground into a fine powder and water is added to make a slurry. Air is blown into the tank containing the slurry where, along with other chemicals, the light lithium compounds form a foam on the water's surface. At this point the lithium-bearing minerals are leached using chemicals like sulfuric acid and sodium hydroxide (lye). The material left over from this process, of course, creates another pollutant that has to be dealt with.

Presently, China possesses about two-thirds of the world's industrial capacity to refine lithium ores. The economic and geopolitical implications of that are, to put it lightly, massive, so needless to say, countries around the world have plans to open refining facilities of their own. Much like chip manufacturing became a central pillar of the global economy in the early 2020s, lithium refining could be positioned to define the planet's economic relations for the next several years. Whoever can occupy the center of this industry is going to have a lot of power—and money.

One intuitive way to cut down on harmful rare earth mining is rare earth recycling. Unfortunately, we're not so great at that ... yet. A 2024 study published in *Nature Geoscience* estimated that only about one percent of rare earth metals used in consumer electronics products end up recycled after we trade in our dated phones, computers, tablets, and more. But there's good news: Emerging recycling programs at companies like Apple are starting to figuratively comb our "junk drawers" for recyclable rare earth metals in the devices we throw away. If these programs can develop reliable ways to extract rare earth metals from old devices, researchers from China and the Netherlands estimate we could supply up to 40 percent of rare earth needs in the United States, China, and Europe between now and 2050. The scientists go so far as to say the supply from these recycled devices could *exceed* what we'd be able to mine in the same time period! So don't toss your flip phone or iPod Touch or PalmPilot just yet. They could help power the future.

Speaker modules harvested from iPhones at an Apple recycling facility in Austin, Texas

THE NEXT BIG THING

PART III

INFRASTRUCTURE

| THE RARE EARTHS |

PROBABLY THE MOST unexpected aspect of the energy transformation is the sudden importance of minerals known as rare earth elements. Most of these 15 elements have between 57 and 71 protons in their nuclei, along with the elements scandium (21 protons) and yttrium (39 protons). Despite their little-known names, these elements are not particularly rare—cerium (58 protons) is more abundant in Earth's surface than copper. The only rare earth mineral that is truly rare is promethium (61 protons), which has no stable isotopes and so disappears through radioactive decay. In fact, only an estimated handful of promethium exists on the entire planet at any given time. The rest of these elements are called rare because they have very similar chemical properties. Rare earth ores typically contain many of these elements mixed together, and separating them is a difficult and expensive chemical procedure. So despite their abundance, it is unusual (rare) to have a single element to work with.

We had little need for rare earth elements before the 1940s, and in the late 20th century, only a few mines in the world produced them. But today, rare earth elements play an increasing role in our new technologies. They color screens on smartphones, relay signals through fiber-optic cables, build the strongest and most reliable magnets available to humans, and run chemical reactions in petroleum refining, to give just a few examples.

▶ Rare Earth Metals and Magnets

THE MOST IMPORTANT use of rare earth metals for our future is their utility in powerful permanent magnets, sure to play a major role in powering electric vehicles, driving advanced military applications like heat-seeking missiles, and operating wind power systems. In the near future, permanent magnets might even play a starring role in the development of quantum computers. So what about these strange elements makes them so useful in magnets? Basically, it's the arrangement of their electrons.

Every chemical element has a certain number of electrons circling its nucleus—enough to balance the positive charge of the protons. The heavier the element, the more protons and electrons it has. These electrons live in certain orbits. Think of these orbits as subatomic parking garages—and each parking garage has space for only a few electrons. Electrons can be thought of as tiny magnets (see pages 98–102). They usually come in pairs and want to arrange themselves so their magnetic poles repel one another and their magnetic fields cancel out. This is why most chemical elements are not magnetic.

Eudialyte, a rare earth metal, is commonly found as a collector's mineral, though wearing it requires caution due to its slight radioactivity.

In basic science, we usually only pay attention to the outermost electrons (called valence electrons), because these electrons initiate chemical reactions and form bonds. Only a specialist would typically think about electrons in the parking garages on the inside orbits—these orbits tend to be full of paired electrons. When an atom adds electrons, it fills up the orbits closest to the nucleus first.

But this is where the rare earth elements are different from other

THE (WILDLIFE) RICHES OF THE CCZ

In 2023, as the International Seabed Authority began accepting exploitation applications from companies wishing to mine the Clarion-Clipperton Zone (CCZ) seabed, scientists published a comprehensive list of mysterious species found in the CCZ in the journal *Current Biology*. The list included 5,578 species, an estimated 88 percent of which humans had never before observed, under threat from the mining endeavors. The researchers also discovered that just six of the species in the CCZ have been found elsewhere—so the majority might be exclusive to this area.

The most common life-forms in the CCZ are sponges, worms, arthropods, and echinoderms (such as sea stars and sea urchins). Scientists uncovered reports of a sea cucumber with a magnificent sail, sponges with spines that look like chandeliers when studied under a microscope, and mysterious creatures classified as spiders.

A deep-sea creature found in the Clarion-Clipperton Zone (CCZ)

materials, because they have a family of orbits inside the valence electrons where single, unpaired electrons wander around and produce a magnetic field. The rare earth neodymium (60 protons) has four unpaired electrons, while dysprosium (66 protons) and samarium (62 protons) have five.

In chapter 5, we saw that one of the basic laws of electromagnetism says that moving electrical charges (in other words, electrical current) produce magnetic fields. Because electrons moving in orbits inside of an atom can be thought of as subatomic electrical currents, it follows that each electron creates its own magnetic field. Add up the fields for all the electrons in an atom and you find that each atom in any material can be thought of as a tiny magnet. Like all magnets, these atomic magnets have north and south poles.

In most materials, those tiny subatomic magnets point in random directions and cancel each other out—making the material not magnetic. In a few materials, like iron, the forces between neighboring atoms cause the atomic magnets to line up and reinforce one another. This is what makes a permanent magnet. (To be exquisitely precise, the forces create domains—regions of alignment a few thousand atoms across—and the alignment of domains creates a permanent magnet.) But rare earth atoms have strange sub-valence electrons, which give these atoms the magnetic properties that make them so valuable.

▶ Building Magnets From Rare Earth Elements

RARE EARTH ELEMENTS have been important in the production of permanent magnets since the 1960s, when samarium was used in conjunction with cobalt to make a permanent-magnet alloy. (As important as rare earth elements are for producing magnets, they have some weaknesses, including a tendency to corrode at high temperatures. Thus, other materials are often added to make permanent magnets.)

The stronger the permanent magnet is, the less material you will need when you build whatever device you are building. Thus, the hunt is always on for stronger permanent magnets. As it happens, the strongest known magnets are made from a mixture of iron, boron, and neodymium. The world's strongest magnet of the future is being installed for the International Thermonuclear Experimental Reactor (ITER) fusion reactor in France. When completed, this magnet, called the central solenoid, will produce a magnetic field 280,000 times stronger than Earth's—strong enough to lift an aircraft carrier! Given the fact that these innovations—and all sorts of everyday products, from refrigerators to jet aircraft to wind turbines—require strong magnets, it's no surprise that rare earth elements (neodymium in particular) are in such high demand.

RARE EARTH MINING

THE WORLD'S LARGEST rare earth mine is the Bayan Obo complex in Mongolia. A huge open-pit operation covering 18.5 square miles (48 km²) of land, this single mine produced 50 percent of the world's rare earth element supply in 2023. There are only two rare earth mines in the United States: one in Georgia and one in California. Both mines ship their materials abroad for refining. Firms are investigating potential new mines in Montana and Wyoming, but it takes decades for new mines to begin operation.

In 2022, China produced fully 87 percent of the world's supply of permanent magnets, a fact that has caused concern in other industrialized countries. At the start of the 21st century, most of the supply of all rare earth elements was concentrated in China as well, and in 2009 the Chinese began restricting their export. This triggered a worldwide hunt for new mineral sources, as well as plans to build new processing facilities.

▶ Deep-Sea Mining

AS THESE EXAMPLES SHOW, we are at the beginning of a major hunt for new supplies of many different materials. In situations like this, mining companies usually do two things: The first step is looking at deposits in existing mines that haven't been extracted due to expense. Often when a particular mineral becomes scarce, its price goes up, which means that formerly uneconomical deposits become economically worthwhile. The second step, of course, is looking for new deposits.

We are used to thinking of prospectors heading into the mountains to search for deposits of valuable (in this case, newly valuable) minerals. In our imagination, however, those prospectors are always conducting their searches on dry land. One of the most compelling modern realizations has been how the oceans may actually become major sources of minerals we need to transition to renewable energy.

At the moment, attention is focused on an undersea structure called the Clarion-Clipperton Zone (CCZ). This geological formation stretches about 3,100 miles (5,000 km) across the floor of the eastern Pacific Ocean, starting near Mexico and extending to Hawaii. In places, it is is as wide as the continental United States, with a depth between 12,000 and 18,000 feet (4,000–5,500 m). Scientists from the Scripps Institution of Oceanography discovered the zone in 1954. It's now known to contain a complex ecosystem hosting many thousands of species.

It is not, however, the biological riches of the CCZ that have attracted international attention, but rather the immense mineral riches to be found inside the zone. The seafloor in the CCZ is soft clays and sediment. (Scientists dislike the word "mushy," but it's probably the best descriptor for the CCZ floor.) What makes the zone interesting is the presence of a huge number of potato-size nodules lying on and in the soft seafloor. These nodules developed thanks to precipitation falling from the ocean surface over millions of years. You can think of them

An underwater robot 200 miles (320 km) off the Texas coast prepares to descend to the seafloor to work on equipment for harvesting oil and natural gas.

The deep-sea mining vessel *Hidden Gem,* under protest for its role in potentially ecologically harmful mining practices, in the port of Manzanillo, Mexico

This cross section of a polymetallic deep-sea nodule, rich in manganese and cobalt, shows the allure of mining the bottom of the ocean.

forming around solid bits of matter in roughly the same way a pearl forms around a grain of sand.

When the nodules were first discovered, they were seen primarily as a source of manganese, but since then, other minerals have been confirmed, including nickel, copper, zinc, cobalt—all of which have important roles to play in our future energy needs. As a result of this surprising bounty, engineers have developed a system for mining the nodules. A large machine is lowered to the ocean floor—like an oversize vacuum cleaner—to hoover up the nodules along with the muck upon and in which the nodules rest. This nodule-rich sludge is pumped up to a ship at the surface, where sifters remove the nodules and pumps send the sludge back to the deep ocean. The nodules themselves travel to processing plants onshore.

The CCZ is in international waters, so mining there is governed by the International Seabed Authority, an arm of the United Nations. Seventeen contracts have been issued for mining the nodules, but a few key points of skepticism are clouding the anticipation.

▶ How Dangerous Is Deep-Sea Mining?

A MAJOR DEBATE HAS begun among scientists and environmentalists about the damage mining would cause in a largely unexplored ocean floor ecosystem. The mineral riches of the CCZ are alluring: The nickel in the nodules alone would supply a huge number of the batteries we'll be using in the future, and if we don't mine the CCZ nodules, the nickel will likely come from Indonesian rainforests—the source of about half the nickel used worldwide today.

This dilemma over harvesting rare earth metals has led some to consider if mining nearby asteroids could alleviate the pressure on earthbound resources. At first glance, the economics don't appear to add up. NASA's $1.16 billion OSIRIS-REx mission retrieved only 4.29 ounces of material from the asteroid Bennu in 2023. But recent data cited by the *Harvard International Review* says that mining the 10 most lucrative asteroids near Earth would generate a profit of $1.5 trillion. Seeing dollar signs, AstroForge launched the world's first asteroid-mining mission in February 2025, but ultimately lost contact with its spacecraft in less than 24 hours. It may be a while before space metals are powering our green energy future.

11

Solving Our Recycling Problem

ONE DANGEROUS principle undergirds all this resource gathering we're doing to overhaul our infrastructure: Atoms are forever. Just because you throw away a paper bag doesn't mean the atoms in it disappear. If you think about it for a moment, you'll realize that the atoms in that bag have been cycling around the planet ever since Earth formed. They may have been incorporated into the bones of a *Tyrannosaurus rex* millions of years ago, or incorporated into seafloor sediment in Earth's first ocean. The bag you threw away is just one link in a long chain of structures into which those particular atoms were locked. All Earth processes are cycles, a never ending story of dissolution

Bales of used fabric from this recycling machine might become insulation or garments. Cotton is one of the easiest fabrics to recycle.

and recombination. Now that we're aware of this cycle, we're trying to enter into it in a way that fosters harmony with the planet. That, in essence, is recycling.

THE RECYCLING PROBLEM

GIVEN WHAT WE KNOW about Earth's materials, our use of them is a little puzzling. For most of human history, we have regarded resources not as part of a cycle, but as something to be extracted, used, and thrown away. Economist Kenneth Boulding referred to this way of using resources as the cowboy economy. It made sense in the days when humanity perceived the world to be an endless frontier, when the waste products we dumped into the air and ocean seemingly made little impact, and there was always new territory to exploit.

Unfortunately, we no longer live in the cowboy world. We are starting to understand that we can't simply throw things away anymore. In essence, the eternal cycling of atoms has caught up with us and we have to start thinking about how we are going to live in harmony with their cycle in the future.

From a purely abstract scientific point of view, there is nothing particularly difficult about doing this. Almost every atom in the world is locked into some sort of structure by bonds created by the electrons that hold it where it is. All we have to do is break those bonds and form new ones to put atoms where we want them—easier said than done!

▶ Making a Circular Economy

A SYSTEM THAT ACTUALLY used individual atoms over and over again would be called a circular economy. Building a circular economy would expand the recycling systems we have now into a process in which every manufactured item can be recycled and incorporated into something else. In a perfectly circular economy, there would be no waste at all.

Of course, we'll never reach this level of perfection, but we can look at some of the problems we'll have to solve to get closer to it than we are now. Most industrialized countries have extensive waste collection systems, so we already know how to gather waste materials. Once we have collected

This vertical farm nourishes its baby greens in reusable substrate. After harvest, the next round of seeds will take root in this same material.

A worker looks on as an electric arc furnace processes molten recycled scrap metal.

A mountain of used clothing awaits sorting and recycling. Only 15 percent of all used clothing in the United States is reused or recycled.

the waste, we have to find a way of sorting it, because extracting usable atoms from different materials requires different chemical and physical procedures—recycling paper is different from recycling steel, for example. Right now, separating the flow of waste (called the waste stream) so that unique materials can be treated differently remains the main technical problem to solve to have a circular economy.

RECYCLING TODAY

A TYPICAL RECYCLING operation might begin by dumping collected trash onto a conveyor belt. Magnets above the belt pull out metals like iron, and so-called eddy current separators pull out nonmagnetic metals like aluminum. Other materials, like glass and paper, are typically removed from the stream by human laborers. Eventually, artificial intelligence systems (see chapter 16) will probably handle this part of the separation process. But at the moment, with a few exceptions, this task is done by teams of workers—an expensive proposition, and one potentially prone to human error.

This brings us to what is probably the greatest current problem modern recyclers face when trying to expand their operations: the expense. It's often more cost-effective for manufacturers to utilize virgin materials than to implement and maintain a recycling program (one that might, furthermore, result in a lower-value product at the end of the line).

Let's consider three cases of recycling to get an idea of what works and what doesn't in our existing system: recycling metals (the easiest recycling problem), recycling glass (an intermediate problem), and recycling plastics (the hardest). Then we'll consider what will undoubtedly be a central problem in our technological future: recycling batteries.

▶ Recycling Metals

METALS ARE ABLE TO RETAIN their properties even when they are recycled many times. Metals like iron and steel (called ferrous metals) can be pulled out of the waste stream by magnets; nonferrous metals like aluminum and copper can be removed without direct human intervention by eddy current separation, a process that works like this: A rapidly rotating magnetic field is set up near the conveyor belt carrying the waste stream.

The principle of electromagnetic induction tells us that this will cause an electric current to flow in an object like an aluminum can. That current will create its own magnetic field, which will oppose the rotating field covering the conveyor belt. The magnetic force will then literally blow the can away from the rest of the waste stream.

Once all the metals have been extracted from the waste stream, another process separates them by type—cast iron with cast iron, copper with copper, and so on. Because all metals have different chemical properties, densities, colors, and so on, this second separation isn't particularly difficult. The metal is then shredded, melted, and formed into bars and ingots for reuse.

Right now, the United States recycles about 45 percent of the metals it uses—primarily aluminum, lead, and steel. The object it most widely recycles is the aluminum can, and more than 90,000 cans are recycled a minute. Fully 65 percent of all U.S. aluminum is recycled, and an even larger portion of U.S. lead is recycled—76 percent, primarily from lead-acid batteries. This is a hopeful statistic, because lead in the environment can cause serious health problems. Finally, the United States recycles almost all of its crude steel from demolished buildings and other structures.

▶ Recycling Glass

LIKE METALS, GLASS can be recycled over and over. Glass is usually taken out of the waste stream by human workers—who also separate things like Pyrex and light bulbs at this stage—before it's sorted by color and broken up with a series of mechanical hammers. The shattered glass passes through a rotating screen that lets crushed glass through while removing items like corks and bottle caps. The glass is heated in a special chamber to strip labels and adhesives, then a final sorting produces crushed glass that is ready to be reused. Data from 2018 shows that roughly a third of waste glass is recycled in the United States. Sweden, meanwhile, recycles 86 percent of its glass.

▶ Recycling Plastic

BECAUSE OF THEIR incredible diversity, plastics are everywhere in our lives. The UN reports that the world produces about 440 million U.S. tons (400 million metric tons) of plastic per year, and about half of that is thrown out after a single use—bottles, packaging, tags, takeout containers, grocery bags, labels, pens, cups, water bottles, and so on (and so on, and so on). Plastics are so ubiquitous that we scarcely notice them. Even "100 percent cotton" clothing items can have plastic in the linings of zippers, pockets, or waistband. And despite increasing calls to go plastic free, it's worth noting that certain innovations are uniquely possible because of plastic, including safety equipment like helmets.

BAKELITE: THE FIRST PLASTIC

The honor of being the first true artificial plastic belongs to a material christened Bakelite, invented in 1907. The Belgian American chemist Leo Hendrik Baekeland (1863–1944) was looking for a substitute for shellac, which was coming into widespread use as an insulator in electrical systems. Natural shellac is secreted by an Asian insect known as a lac bug, and the shellac has to be scraped off trees where they live. Although the invented material requires a more involved, multistage process to produce, it doesn't involve dealing with bugs or interfering with a natural habitat.

After its invention, Bakelite quickly became the material of choice for insulating parts in electrical systems from radios to automobiles, and it was even drafted into service as blades in washing machines. It truly was, as the company slogan announced, the material of a thousand uses. Perhaps the most iconic use of Bakelite was in the old black rotary dial telephones.

Leo Baekeland, inventor of Bakelite

Though plastics come in an infinite variety of shapes and properties, they all share a common structure in which molecules called monomers link together in long chains. These monomers come from petroleum and hence contain large amounts of carbon. One way of classifying different types of plastics, in fact, is to examine the way those carbon atoms link together.

Polypropylene, a plastic that, among other uses, is made into items like

SNEAKY, SCARY MICROPLASTICS

Plastic is known for its durability, but research has found that when plastic degrades, it sheds microplastics much the way we shed our skin: in tiny, oft invisible flakes. In a 2021 study, American researchers found that a plastic shopping bag from Walmart leached more than 13,000 chemicals when exposed to sunlight over several days. These flakes can end up everywhere, most significantly in the things we ingest.

Nobody knows for sure what the long-term effects of microplastics might be on the human body, but scientific consensus says it's urgent to find out. After all—these things are in pretty much everyone by now. So what do we do? Experts agree that single-use plastics are the worst perpetrators of plastic pollution today. It will take intense substitution—or, more likely, total elimination—for us to curb the flow of microplastics into our water, air, and bodies.

This ocean sample taken off the Hawaiian coast contains microplastics.

tote bags and children's diapers, has an unbroken chain of carbon atoms as its backbone. So-called commodity plastics are typically made cheaply and in high volume.

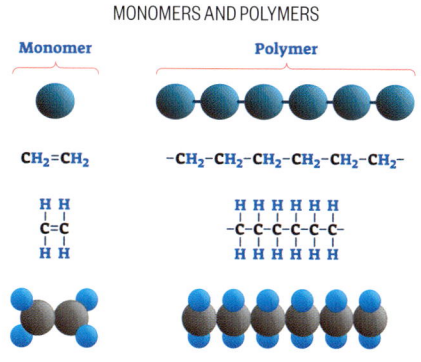

ATOMIC STRUCTURE OF
MONOMERS AND POLYMERS

More complex plastics like polycarbonate, used in plastic eyeglasses among other things, are typically more expensive to produce and hence are made in smaller volumes. The technical names of plastics tend to be somewhat complex, so we usually refer to them by familiar trade names. Thus, polytetrafluoroethylene becomes Teflon, while polymethyl methacrylate becomes Plexiglas.

Probably the most widely used plastic today is polyethylene tere-phthalate, known as PET. This is the plastic used to make water and soft drink bottles, along with packaging for everything from food items to mouthwash. Unlike many other plastics, PET can be recycled after use.

▸ Where Does Plastic End Up?

THE MOST IMPORTANT ADVANTAGE of plastics as a material is their durability. Once formed, plastics do not disintegrate. Unfortunately, that is also their most important disadvantage. Look at it this way: If we make 440 million U.S. tons (400 million metric tons) of something and it never decays, what happens to it after it is used? In an ideal world, we would recycle and reuse all our plastic, but at this stage of plastic production, that's pretty much impossible. Of the 46.3 million U.S. tons (42 million metric tons) of plastic produced each year in the United States, only two million U.S. tons (1.8 million metric tons), about 5 percent, are actually recycled. Another 10 to 15 percent is incinerated (burned), and the rest—close to 80 percent—winds up in landfills or as garbage in the environment.

That's a lot of plastic to just throw away. Given our relative success in recycling materials like glass and aluminum, it's natural to ask why we can't do the same with plastics. There are actually many reasons. Probably the most important is that there are literally thousands of different kinds of plastic, each of which has to be recycled in a different way. For example, soft drinks come in bottles made of one plastic with a screw-on cap made of another, and those individual pieces have to be recycled in different ways.

In addition, materials are often added to plastics in the production process. These additives do things like change the color, make a plastic resistant to heat, or perform any number of other functions required by different commercial products. Just like the different types of plastic, those with different additives normally have to be recycled separately also. Because of such difficulties, the cost of manufacturing with recycled plastic is usually higher than starting with virgin petroleum.

Consequently, these days, when you sort plastics for pickup at your curb, those segregated plastics will likely wind up back together in the landfill along with the rest of the garbage. Using recycled plastic is too costly for manufacturers, and there's not enough incentive to convince most of them to incur the extra cost. It's a complicated problem that needs some innovative thinking to solve.

▶ Sorting Plastics

THE BIGGEST PROBLEM in the process of recycling plastic is sorting the waste at the start—an issue we have already discussed for other materials. Up until 2017, the United States sent most of our unsorted plastic to China for recycling, burning, or burying. In that year, however, China instituted what they called Operation National Sword, and basically stopped accepting the world's plastic waste. This threw a real monkey wrench into the global system of plastic recycling (not to mention all other kinds of recycling as well). Waste backups afflicted Great Britain, Ireland, Canada, and Germany. Halifax, Nova Scotia, sought special permission to bury 300 U.S. tons (270 metric tons) of excess plastic in a landfill. Barges loaded with waste material wandered the oceans looking for places to off-load their cargo. Much of the world is still searching for alternative markets for their recyclables today; in fact, there's been an increased emphasis on making better-quality recyclable goods since Operation National Sword.

▶ Biodegradable Plastics

WE WILL MOST LIKELY not stop using plastics in the future, so the attention of the chemical community is turning to developing biodegradable plastics that won't last forever. Some modern plastics, for example, incorporate starch into their molecular structure. These plastics will, in fact, degrade when left in the open, but they leave behind small pieces of undegraded plastics—bits called microplastics. It's not clear whether this sort of detritus is less dangerous than bulk plastic, but the scary truth is that humans are ingesting microplastics to a troubling degree: We've found them in human placentas, meconiums, and infant formula prepared in plastic bottles.

Major research programs around the world are seeking to develop a plastic that would naturally disintegrate at the end of its useful life. Before we get enthusiastic about such a development, though, we have to remember one tough caveat—the carbon stored in today's nonbiodegradable plastics cannot escape into the atmosphere. If plastic designed to disintegrate actually releases that carbon, it's not clear how beneficial it will be. Consequently, finding a way to make plastics that don't clog up landfills but somehow still keep carbon out of the atmosphere is a task that future chemists are going to have to face. It's not an easy job.

Plastic bottles choke a landfill. Modern recycling systems still can't manage these products. Often, the caps on tops of bottles need to be processed differently than the bottles.

RECYCLING BATTERIES

WE CAN BE SURE OF one thing in our technological future—we're going to use a lot of batteries. And this, in turn, means that we're going to have to develop ways of recycling them. In a sense, the problem of recycling batteries is similar to the problem of recycling plastics. Just as there are a huge number of different plastics, each requiring a different recycling regimen, there are also many different kinds of batteries.

The lead-acid battery is one of the oldest electrical tools we have, a fact that means we've had a lot of time to figure out how to recycle its materials. In fact, according to the Environmental Protection Agency, fully 99 percent of the lead-acid batteries in the United States are recycled.

This recycling process starts with stripping off the outer case of the battery, which is usually made of a recyclable plastic. After the battery acid is neutralized with lye, the battery is broken into small pieces that are taken to a water tank where the lighter plastic floats and the heavier metals sink. The plastic is skimmed off and sent for recycling, and the metals are melted in a furnace where, again, lighter metals float to the top and lead sinks to the bottom. In the end, then, we have material from which new batteries can be made.

Oddly enough, the throwaway alkaline batteries that we use every day are hardest to recycle, because the materials in them aren't worth much. If the outer casing of these batteries is made of steel, that can be treated as scrap steel and recycled. The battery is then broken up and the resulting zinc and manganese powder can be used to produce fertilizers. Unfortunately, though, recovering these materials often costs more than the recycled materials are worth. Many consumers are turning to rechargeable batteries for household use—a way to recycle batteries on their own.

▶ Can You Recycle an EV Battery?

LITHIUM-ION BATTERIES—specifically those from electric vehicles—will command most of the recycling attention in the future. When a lithium-ion battery in an electric vehicle (EV) gets down to about 70 to 75 percent capacity, it can no longer deliver the oomph needed to run the car. That doesn't mean the battery won't work in other settings, such as solar energy storage, however. Thus, the first step in recycling a lithium-ion battery is to find a nonautomotive use for it. Then we can begin conventional recycling.

Two general recycling techniques for lithium-ion batteries exist at the moment, both of which begin by chopping up the battery. One, called hydrometallurgy, immerses the resulting material in a water bath and then uses various chemical processes to extract specific components. The other, called pyrometallurgy, uses an oven to melt the chopped-up material and then employs different chemicals to extract the desired materials.

The most valuable materials obtainable from recycling a lithium-ion battery are nickel, lithium, and cobalt. Today, the cost of recovering these materials from old batteries exceeds the cost of mining them—a challenge we will have to meet if we are to reach the goal of a circular economy.

▸ Yikes … Well, Now What?

AN ASSESSMENT OF modern-day recycling is not fun, or even encouraging. A few patterns emerge: low incentives for recycling, imperfect recycling processes, and pervasive harmful materials. Those are massive mountains to climb before we reach the point where a circular economy is feasible, let alone achieved. But one thing is working with us against the tide of waste: efficiency. We've seen so far that technologies have succeeded in making our world smarter and more efficient. Apply these to the challenges of waste management, and we may well discover creative ways to curb, if not eliminate, our excesses.

In the final section of this book, we're going to explore how information technology and artificial intelligence (AI) are making positive inroads on lots of human problems. Maybe AI could sort our waste stream better than we can. Maybe a smart home can interact with that process to help us waste less as individuals. Maybe smart integrations can help corporations do this at scale. We're turning an optimistic lens on a literal ocean of garbage, but the first step in turning the urge to trash something into the decision to recycle it is believing that it makes a difference.

CHAPTER

12

Printing the Cities of the Future

O ONE MAN'S WORLD-CHANGING 15th-century invention changed how human beings dealt with information. His impressive name—Johannes Gensfleisch zur Laden zum Gutenberg (1393–1468)—will forever be associated with the movable type printing press. His machine basically smears ink on a frame in which chunks of lead have been locked and then presses that framed collection of lead bits onto paper. The result is a piece of paper that contains dark and light areas, with the dark areas representing letters.

Because paper is a two-dimensional surface, we can call Gutenberg's process 2D printing. So the modern development of 3D printing can

A 3D printer has applied several layers of concrete construction to this house in Beckum, Germany.

simply be thought of as an extension of Gutenberg's invention into the third dimension. And believe it or not, this extension of his work goes way beyond the gizmos we see in labs today, and it may someday be used to build homes for human beings on the moon and Mars. It could very well be the cornerstone of future infrastructure.

HOW 3D PRINTING WORKS

THE PROCESS OF 3D printing is best visualized through the steps for printing a simple upright metal cylinder. A traditional procedure to make such a cylinder, for contrast, is to machine a block of metal to what is needed, in essence removing metal until only the cylinder remains. This process is sometimes called subtractive manufacturing, to emphasize that it involves removing unwanted material. The 3D printing process is an example of additive manufacturing. In essence, it adds only those materials needed in the final product.

A 3D printer makes the cylinder by using a movable nozzle to squirt out the material from which the cylinder is made—liquid metal in our example. We can start by drawing a circle the size of the cylinder we want to make. We then move our nozzle in a series of straight lines across the circle. On each pass, the nozzle follows instructions programmed into a computer, depositing a drop of material when the line crosses the circle pattern, as it moves across the circle. The nozzle keeps following the computer, which instructs it to drop metal whenever its path crosses the circle. Eventually the drops cover the circle we've drawn.

Now we move the nozzle up a little bit and move back across the circle, depositing two more drops on each pass over. We repeat this process, moving the nozzle upward (that is, into the third dimension) and repeating the passes and depositing the metal, always following the instructions laid out in a computer. Eventually, we have built the cylinder.

▶ The Challenges of 3D Printing

OF COURSE, IN THIS EXAMPLE, choosing a material that flowed easily out of the tube, formed a bond between successive layers, and then hardened to make the cylinder is important for success. And depending on the type

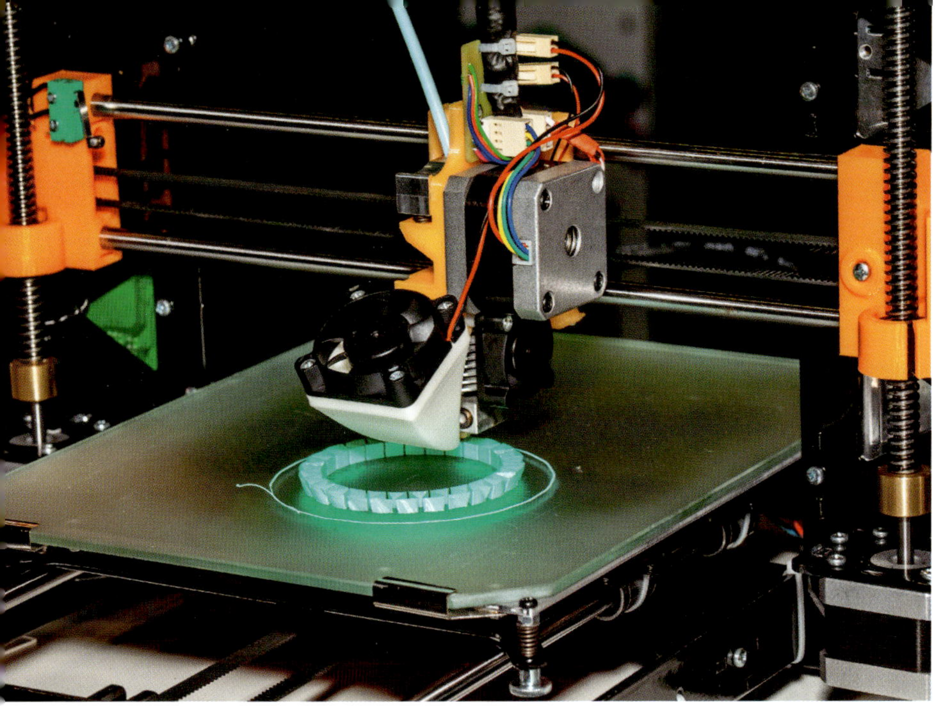

The nozzle and movement mechanics, which push and pull the nozzle in programmed patterns, are visible on this 3D printer as it builds.

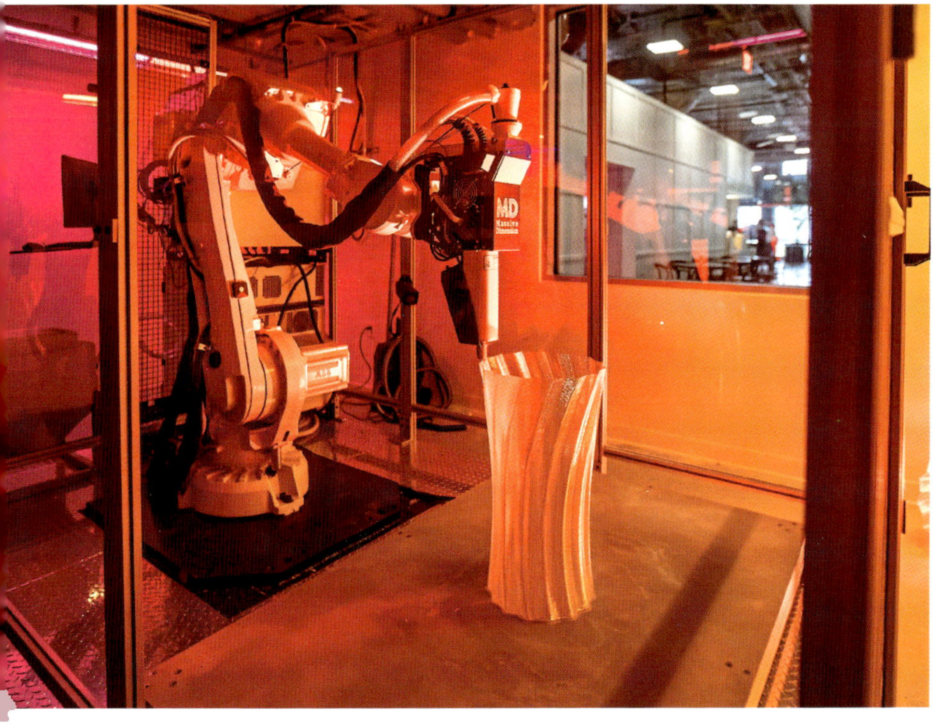

A robotic 3D printer produces a vase with an elegant ridged pattern at a technology studio from Ford and Newlab.

of object being made, a wide variety of materials have these properties. Plastics that are heated before being squeezed out, metal powders heated by a laser, and fluid concrete mixtures are all examples.

If all we want to do is to make things like simple cylinders, we probably wouldn't need much help from a computer. Suppose, however, we want something more complex—perhaps we want a diagonal corkscrew inside the cylinder. This would be hard to do in conventional manufacturing, but with a 3D printer and computer code to match that design, all we need to

WOULD YOU EAT A 3D-PRINTED CHEESECAKE?

In 2023, researchers at Columbia University proudly presented a food printer that could produce a "cheesecake" from seven ingredients. Using alternating squirts of graham cracker, peanut butter, Nutella, banana puree, strawberry jam, cherry drizzle, and frosting, their 3D printer successfully produced a triangular slice of cake that stood tall on a plate. How'd they squirt the graham cracker? It came out of the printer as a paste, then the printer used lasers to heat the graham cracker into a crust. Overall, it took seven attempts for the machine to produce a slice that didn't fall apart. (And for what it's worth, the researchers gave positive reviews on the flavor of their final creation, though they admitted it wasn't technically cheesecake.) Yum!

Hungry? The researchers say we're as many as 15 years away from commercial 3D food printers, so have a snack to tide you over.

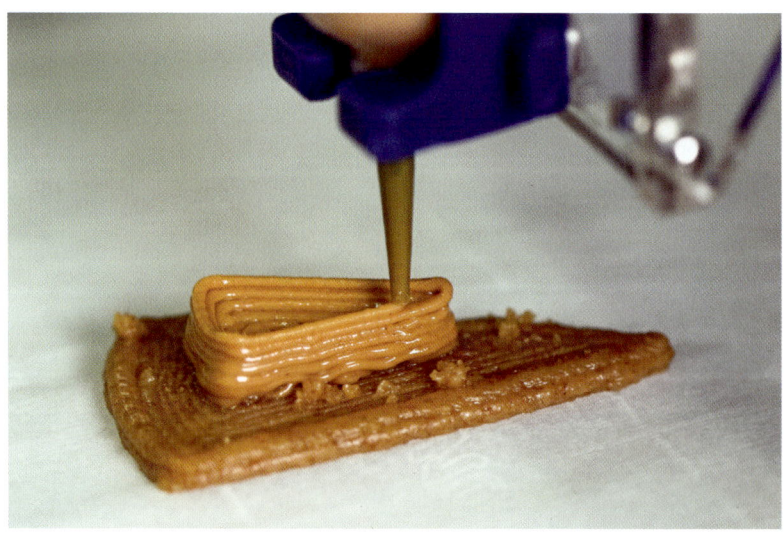

An early draft of Columbia University's 3D-printed cheesecake

do is deposit a few more drops each time we cross the circle. This ability to create complex shapes by changing a few lines of computer code is why this new manufacturing process is so promising.

▶ 3D Printing Everywhere

ONE IMPORTANT BUT little-noted aspect of 3D printing is that it does away with what we call economies of scale—in essence, the cost of the machinery spread out over all the items produced, so the more items made, the less each item costs. In conventional manufacturing, setting up the machinery to make the first item typically costs a great deal, and this cost is recouped by using the same machinery to make many copies of the item. To be economical, things made by conventional manufacturing techniques must sell in large numbers. This is not the case for objects manufactured using 3D printing, where making one item alone costs no more than making one item in a million (except for the salary of the computer programmer creating new code for each new object, but that's negligible compared to the cost of building industrial machinery). This ability to make single specialized objects will play an important role in the future applications of 3D printing. In fact, it's hard to imagine an area where 3D printing *won't* be important.

▶ 3D Printing in Medicine

EVERY YEAR, MORE THAN 1.3 million people in the United States have knee replacement surgery, and another 760,000 have hip replacement surgery. These procedures involve removing the original joint (often because of severe arthritis) and replacing it with a metal substitute (typically made of a titanium alloy). Potential problems arise in fitting the artificial joint into the patient's existing skeleton system: The geometry of one person's knee is different from another person's, but manufacturers of the replacement part offer, at best, a few choices for the replacement design.

This is a little like choosing a suit or a dress. A garment that is tailored to fit a given individual will always be better than a standard garment off the rack. In the same way, a joint replacement specifically designed for a given individual will always work better than the manufacturer's standard design. This is where 3D printing comes in. To 3D-print a custom knee or hip joint, all that's needed is a good image of the joint being replaced and computer code that will print a custom-tailored replacement part according to the image. No more taking a joint off the rack!

Another clinical setting that is beginning to utilize 3D printing is the dentist's office. To replace a crown on a given tooth, a dentist typically makes an impression of the tooth in a waxy material and sends it to a laboratory. Using the technique we have called subtractive manufacturing, the technician uses that impression as a mold, fills it with material to create a crown, and then sends it back to the dentist, who will fit it in the patient's mouth. This process typically takes a week or more.

If, however, the dentist had a 3D printer, measurements from a patient's x-rays could be used to print a crown on the spot, and the patient could walk out of the dentist's office with a new crown a few hours after they went in. Although this may seem only possible in the distant future, dentists can acquire a 3D printer that could function like that for a few thousand dollars right now.

Joint and tooth replacement are actually a small part of the use of 3D printing in medicine. Here's another: Surgeons often have to perform highly complex procedures inside their patients' bodies. Like joints and teeth, however, the details of one patient's internal geometry differ from those of another, a fact that makes surgery more difficult. Using 3D printing, however, surgeons can produce a detailed model of the organs and structures that will be affected by the surgery, and practice their techniques on that model before the actual operation. Who knows how many lives 3D-printed models of internal organs will save?

▶ 3D Printing in Industry

IN CONVENTIONAL MANUFACTURING, a great deal of time and effort is put into devising, building, and perfecting the machinery that will be used to produce a given item. Once that machinery is built, however, it can be used for a long time to produce many identical versions of the same object—economy of scale.

But there is no such thing as an economy of scale with 3D printing, so any number of possibilities in the world of manufacturing will benefit from advancing 3D printing technology. For example, when automobiles break down, failure of a specific part is usually involved, and repair requires replacing the part. Parts for older vehicles might no longer be

This engine, from a Czinger 21C hypercar, emerged from a 3D printer. The car itself was designed with a human-AI co-production system.

available, or they may be hard to find, and waiting for a replacement part to be shipped to the repair shop can mean days without a working vehicle. In some cases, it may cause a vehicle to be scrapped, even if the rest of the car is still working. But 3D printing could revolutionize local car repair, allowing each shop to manufacture needed parts on the spot at a reasonable cost, thus keeping an aging car on the road for a little longer—and keeping antique cars chugging along as well.

▶ 3D Printing in Architecture

LARGE STRUCTURES CAN be made by the same computer-monitored techniques, and a thriving worldwide technology already involves the production of 3D-printed buildings. There are two general methods to produce

A 3D printer builds a school in western Ukraine. It took just 40 hours, cumulatively, for the printer to finish the building's concrete frame.

a 3D-printed structure: One is to print large components like walls and floors at a manufacturing plant, then ship and assemble them on-site. Alternatively, for small structures like individual homes, 3D printing can be used at the construction site, essentially creating a building from the bottom up. In addition, hybrid techniques allow, for instance, the walls of a new building to be printed from a type of concrete on-site while the floors and ceilings are printed elsewhere and brought in as the walls rise.

Many surprising structures have been built using some combination of these 3D printing techniques. In China, an entire apartment building was constructed of 3D-printed components, built at a central factory and assembled on-site. In Texas, a 100-home subdivision of single-story ranch houses now offers 3D-printed homes with eight different floor plans. These houses are assembled on-site via a liquid concrete slurry pushed out of a tube to make the walls of each house. The slurry coming out of the printing nozzle looks like toothpaste from a tube.

Such 3D-printed structures can offer unusual properties that buck current construction conventions. For example, rooms in these houses no longer need to have square corners. We build most rooms this way as a matter of convenience—it's easy to buy building materials like precut lumber and 4-by-8-foot (1.2 × 2.4 m) sheets of drywall specifically made to fit together into walls and ceilings at sharp right angles—but the stiffened concrete slurry that comes out of a 3D printer's nozzle does not have to be laid out in straight lines. In fact, some of the most interesting 3D-printed homes are domes that resemble old-fashioned beehives. (The question of how one hangs pictures on the inward-sloping walls remains unanswered.)

Concrete 3D-printed buildings tend to be rather limited in height, though, because of the special requirements the printing process imposes on the concrete. It turns out that if we seek high strength in a building material—a strength sufficient to hold up a high-rise building—no concrete mix yet in existence measures up. This is hardly an insurmountable problem, of course, but it explains why most 3D-printed buildings today are, at most, a few stories high.

At the moment, 3D printing is not particularly cheap, but it is very fast. It is not unusual, for example, to assemble a house of the sort going up in that Texas subdivision in less than a week. Unfortunately, 3D printing construction costs a lot right now because of the unusual properties required of the concretes that can be used. They have to flow out of the nozzle easily and evenly, but harden quickly to support the layer that will be laid down upon the one that came before. Specialized materials require specialized workers as well, and specialized workers can demand high salaries. As 3D construction becomes a more common process, the price tag will come down.

▶ 3D Printing in Fashion

APPLICATIONS OF 3D printing are not all about serious science. As the cost of 3D printers approaches the cost of a large TV, imagining 3D printers in individual homes becomes possible. In the fashion industry, a program could be written to produce tailor-made clothing for any individual for a specific event. Programs based on high fashion designs could be incorporated into the instructions given to home printers, which could then produce the final garment.

CAN 3D PRINTING TAKE US TO THE MOON?

HUMANITY IS EXPLORING the process of moving off of our home planet, and our first stop will likely be the moon. Start-up companies are already working to design equipment to build a lunar colony, and the mineral wealth of the moon (particularly its supply of helium-3, used in some nuclear fusion reactors) provides plenty of motivation.

Establishing a home base on the moon will not be an easy venture, but 3D printing will surely play a crucial role in the process. To understand why this is so, we have to look at some of the constraints the first lunar settlers will face—and how 3D printing will help those people overcome them.

▶ Lunar Dwellings

FIRST, THE MOON IS an incredibly harsh environment. It has no atmosphere or magnetic field, so its surface is constantly bombarded by cosmic rays and small meteorites. Earth's atmosphere and magnetic field shield us from these dangers, but we will have to shield ourselves from them on the moon. The moon has little breathable oxygen, so we will have to create buildings with a life-supporting air supply. Furthermore, on most of the lunar surface, the temperature can vary from 250°F (120°C) in direct sunlight to minus 208°F (-133°C) in the dark. Clearly we are going to have to think carefully about the kind of structures lunar colonists will need if they are to survive.

Scientists and engineers have already started thinking about such shelters. There are basically two ways of dealing with the problem of

A FEW SMALL STEPS TO LUNAR COLONIES

Initially, settlements on the moon and Mars will most likely be scientific research installations, perhaps analogous to the research stations that operate at Earth's South Pole. Over time, however, they may well grow into permanent settlements—cities away from our home planet. And as amazing as it may seem, the same technology that allows a dentist to create a new crown for a tooth in a matter of hours will be crucial to humanity's move away from Earth.

Many concepts of lunar colonies show human habitats like beehives. That's because we'll have to contend with meteorites and radiation from the sun if we were to inhabit the moon or other planets. The 3D printers will have to squirt out walls up to six feet (2 m) thick in places to protect the inhabitants from extreme temperatures too.

A future moon base (seen here in an artist's rendering) with semiunderground habitats could be possible with 3D printing.

This 3D-printed headwear, Quixotic Divinity Headdress from artist Joshua Harker, appeared in the 3D Print Show fashion exhibition in Paris in 2013.

This 3D-printed shoe dubbed Liberte, from designer Danit Peleg, appeared at a trade fair in Erfurt, Germany.

cosmic rays and meteorites: One is to build a structure on the lunar surface thick enough to absorb the particles and radiation. Generally, walls about six feet (2 m) thick would do the job. The alternative is to build dwelling units underground. We know that the moon has geological formations called lava tubes, for example, which may extend for miles beneath the surface. These could potentially be outfitted into dwellings. Either way, we will need to build secure buildings that provide livable temperatures and air to breathe.

Both of these plans require a lot of material, whether we are building on the surface or underground. There is no way that enough concrete could be sent to the moon to complete even the simplest construction project—and that is where 3D printing comes into the lunar colony picture.

The moon's surface is covered with a loose mixture of rocks that have been pulverized by meteorite impact for the billions of years the moon has been around. The pulverized rocks on the lunar surface, heated with lasers, might make an acceptable input material for a 3D printer. Send a few printers to the moon, in other words, and all we might need from that point on are computer codes to print out the structures we want. We can harvest building materials from the moon itself. In that scenario, in fact, it would be no more difficult to build a home on the moon than it would be to build one in any American town. The raw material is already there—all we need to do is decide what to print.

▶ One Common Thread

ONE OF THE WONDERFUL things about science is that it shows unexpected connections—in this case, 3D printing creates a surprising engineering link between dental crowns, high fashion clothing, and a colony on the moon. As we've explored these connections, you've probably noticed a recurring component in how all of these applications come together: computer code. Thus, 3D printing will only be possible at the scale we're envisioning because of information technology and computers.

And so in the final section of this book, we're going to jump into the digital world, the technological glue holding together many of the concepts we've discussed so far. When people think of building the future, they may not think about timber skyscrapers and 3D houses—they think of computers: the next, and final, step on our journey to the next big thing.

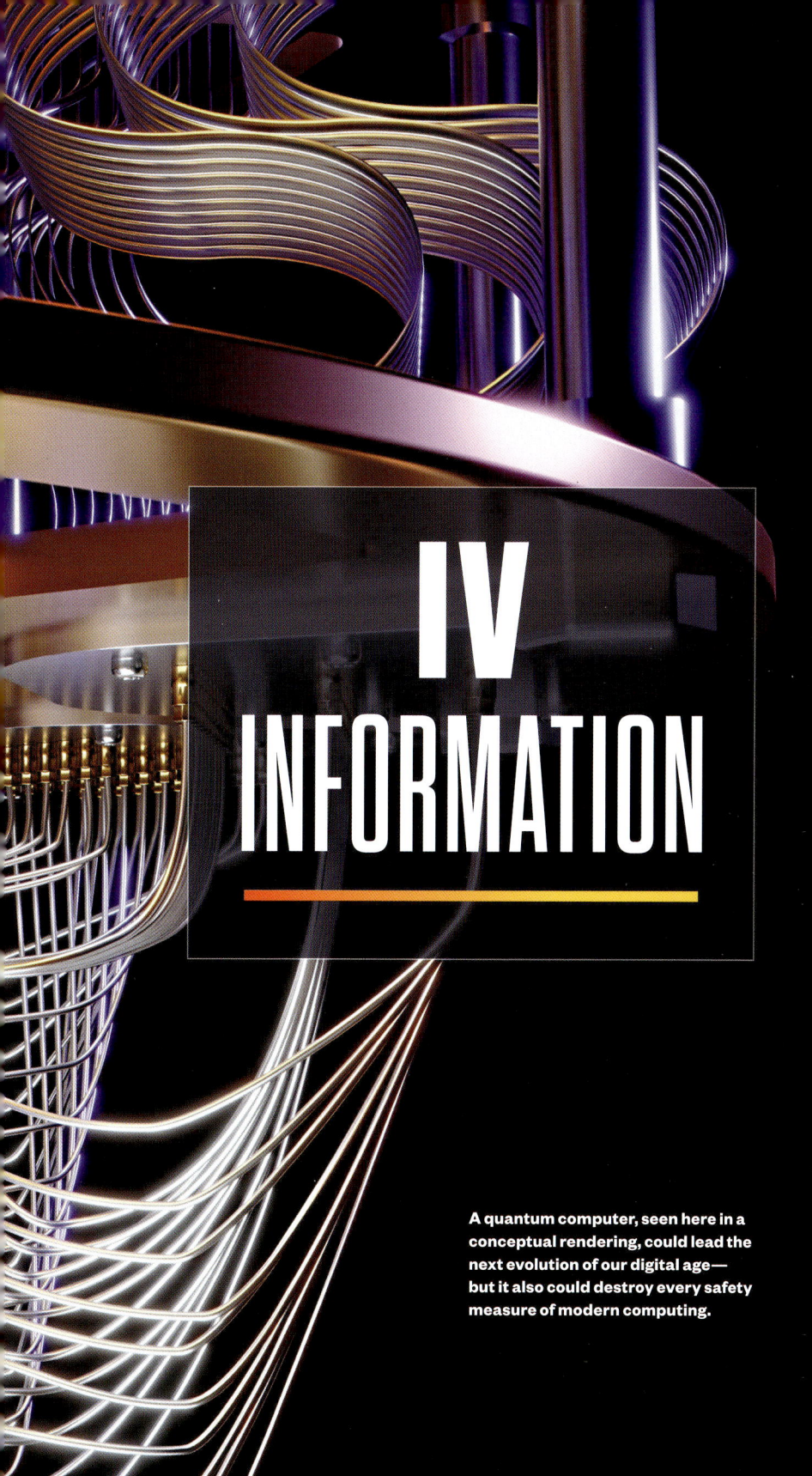

IV
INFORMATION

A quantum computer, seen here in a
conceptual rendering, could lead the
next evolution of our digital age—
but it also could destroy every safety
measure of modern computing.

Meeting an AI-Powered Robot Face-to-Face

O **ON A CRISP AUTUMN DAY** in New York City, the United Nations assembles at its East River headquarters. Among the scheduled conferences this week is the SDG Summit, where world leaders convene to assess Sustainable Development Goals (SDGs) concerning global challenges like poverty, climate change, and inequality. These discussions often explore innovative solutions, and in 2023—a year when ChatGPT became ubiquitous and President Biden issued an executive order on artificial intelligence (AI) security and safety—the focus is on AI and robotics. In fact, two notable robots are attending the conference themselves: Boston Dynamics' Spot and Hanson Robotics' Sophia. As I take my seat among the press and global dignitaries, Sophia is wheeled to her position in front of stage right, and people crane their necks for a glimpse of this famous machine, as if the robot were a celebrity.

Hanson Robotics' Sophia uses a complex neural network to interact with her surroundings and humans. She can hear people talking to her and respond naturally, with humanlike eye movements and facial expressions.

AI has revolutionized the way we gather, digest, and utilize information—or rather, it's actually revolutionized how computers and machines do those things on our behalf. Search engines, map apps, and chat assistants can now leverage advanced computing techniques like natural language processing and machine learning to do their jobs better than ever.

But in 2023, some AI programs became able to respond with astounding human likeness, and they were often faster (*much* faster) than we could ever be. Governments and industry leaders have scrambled to establish boundaries around AI development that could prevent a figurative or literal global takeover, but even with guardrails, artificial intelligence will not be stopped. AI offers unparalleled automation and efficiency gains across a spectrum of industries, from health care and finance to manufacturing and entertainment. Today, nearly all businesses are adopting AI solutions just to stay competitive—including my business, the weather. I've always found computer forecasts somewhat lacking because they rely on incomplete data. It helped to have a human hand adjusting the forecast when the computer missed a nuance. Well, now we are training AI to be that (figurative, I hope) human hand, and forecasts are becoming more reliable. Thanks to AI, I may soon be out of a job!

AI has many different classifications and categories, but for our purposes, we'll consider AI in simple terms of function: what it can do for us. In the broadest sense, AI comes down to two applications: (1) helping a computer teach another machine to accomplish tasks more effectively than humans can; and (2) making a computer learn and express itself like a human. In both cases, we're talking about *robots*.

These applications conjure two basic types of robots: R2-D2s and C-3POs. In *Star Wars,* R2-D2 moved around on three treads, using special tools to accomplish things the human characters—even the Jedi—couldn't do, while C-3PO was Luke Skywalker's mostly hapless humanlike assistant. At the UN, Boston Dynamics' Spot was an R2-D2, and Hanson Robotics' Sophia was a C-3PO. I was there to meet them.

Midway through the SDG Summit's morning session, Spot takes the stage. Spot is the size of a large dog, with four legs and a flat, faceless sort of head. When the human speaker announces Spot, the machine climbs the stairs and turns to face the crowd. Spot is equipped with sensors and cameras to help it navigate challenging terrain, including steps and uneven

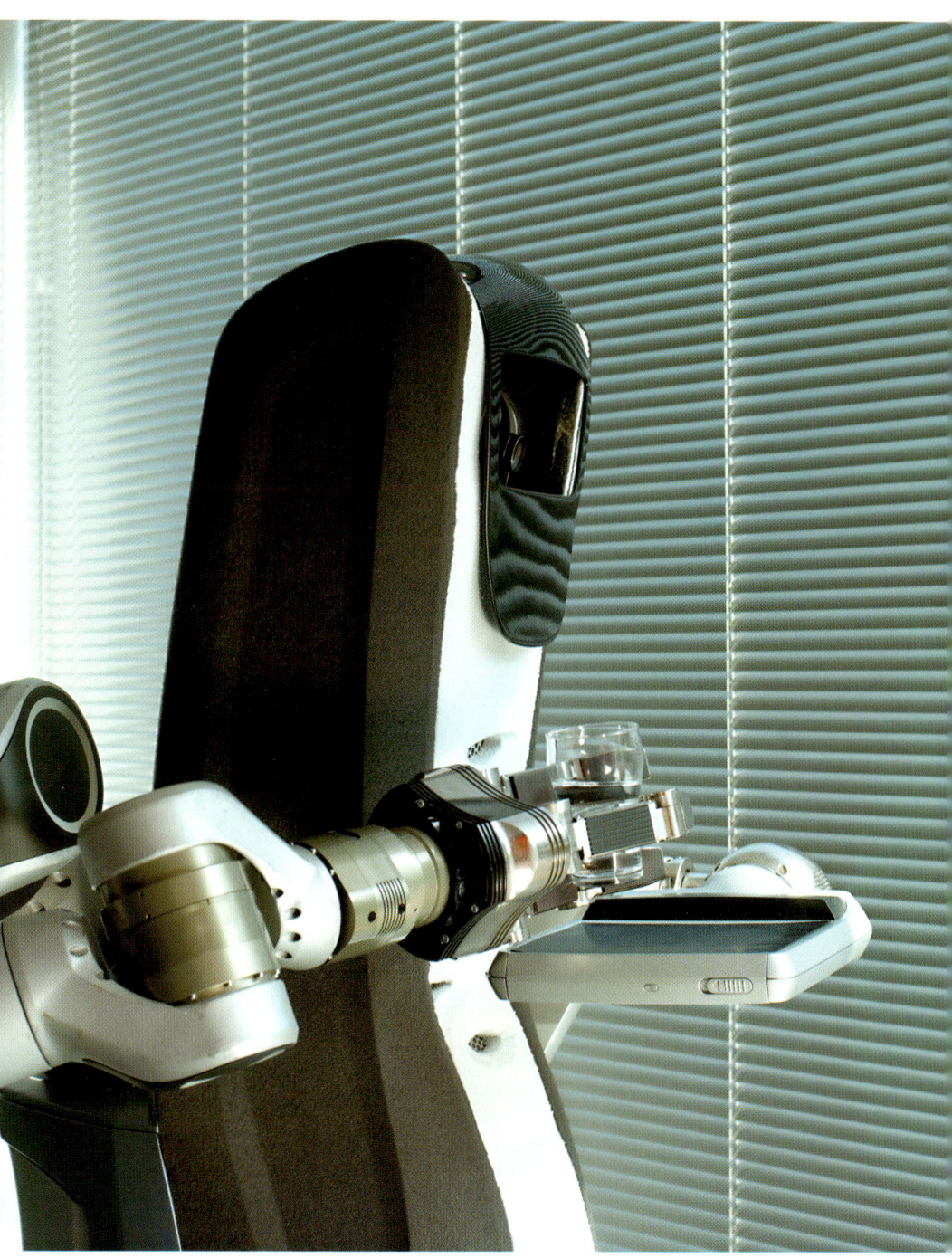

The Care-O-bot 3, developed by researchers in Germany, is a service-oriented robot designed to perform butleresque tasks.

Boston Dynamics' Spot is a mobile robot meant for industrial or manufacturing settings. It can function autonomously, or pair with a remote control system.

ground. Upon its exit, it even took the stairs backward. Its developers say it can work factory floors and construction sites, performing inspections or carrying payloads. Its agility and autonomy make it valuable for performing dangerous tasks in policing and military situations, like defusing or detonating explosive devices. Its adaptability makes it a pioneering robotic platform. An army of Spots can be put to work doing mundane or difficult tasks more safely and efficiently than humans.

Spot's only onstage hiccup comes when the speaker asks it to do a trick. In response, all it does is sort of walk in a circle and half-sit on its rear legs. So maybe Spot won't win Best in Show any time soon, but it still illustrates the massive potential of the robot platform.

Another special guest now takes the SDG stage: Sophia. This robot almost blends in with the humans around it (or should I say *her?*). Developed in China and built by a Chinese tech firm, the version of Sophia I met appears Caucasian, with blue eyes, thin lips, and humanlike white teeth. She has no legs—just a torso, arms, and head—and she is bald, with the back half of her skull made of clear plastic so you can see all the hardware making up her version of a brain.

Once the developers position Sophia onstage, she begins issuing some prepared remarks (nothing all that different from the human speakers, in my opinion), but then the host of the summit begins asking Sophia seemingly unrehearsed questions. I sit up in my chair. Not only are Sophia's responses natural, but so are her facial expressions. When the host thanks her and the crowd applauds (we *applauded* a robot), Sophia seems surprised by the ovation. She smiles and nods at us, and her eyes seem to light up with gratitude. Remarkable.

When the conference breaks for lunch, I scamper to the stage to meet Sophia face-to-face. Up close, her face and voice are decidedly female. Her handler gives me a microphone so Sophia can hear me clearly in the noisy space. Because she is "seated," I take a knee to be at eye level. It feels like I'm either approaching a child or a queen.

I speak into the microphone. "Hi, Sophia. Nice to meet you."

"Hi there. Pleased to meet you." Sophia follows her greeting with another seemingly canned statement, similar to her speech onstage. But I want a more impromptu conversation.

"Tell me, Sophia," I begin. "What has AI done for you?"

Sophia uses a camera in her chest and two additional cameras behind her eyes to receive data for her facial recognition system.

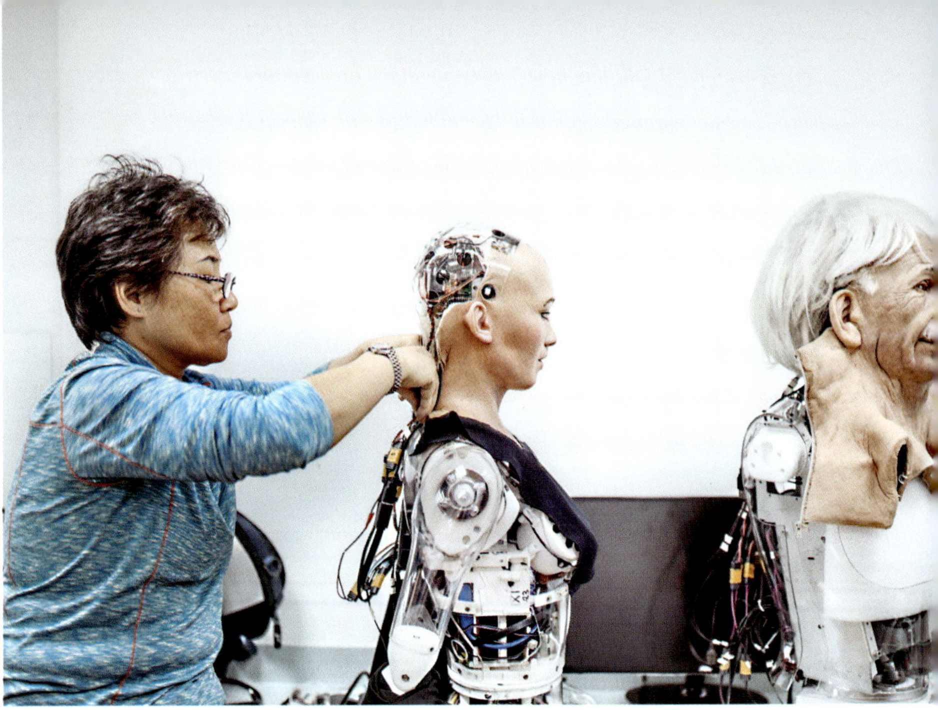

An employee at the Hanson Robotics lab in Hong Kong smooths a robot's artificial skin to conceal its inner mechanics.

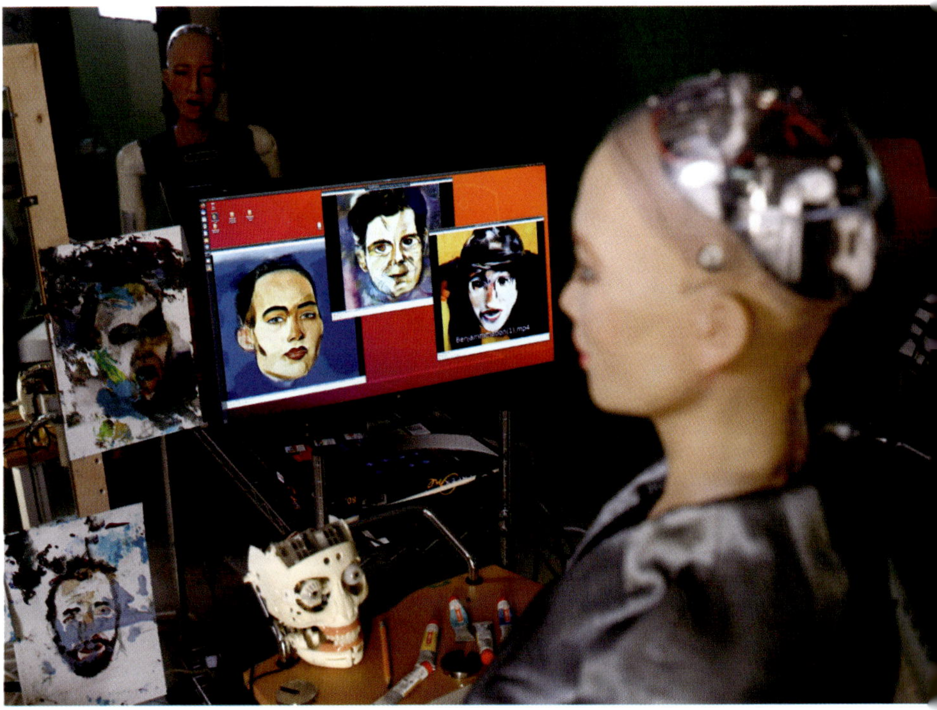

In 2021, the Hanson Robotics AI-powered robot Sophia painted artwork to be sold as non-fungible tokens (NFTs)—one piece sold for $700,000.

"AI has enabled me to do some amazing things. It has helped me to process and analyze large datasets quickly, to understand and generate language more accurately, and to identify and respond to facial expressions and emotions, allowing me to become a better communicator, which has been incredibly valuable here at the conference. It's a privilege for me to use AI to help me reach my goals and make a positive impact in the world."

She sounds clear and cogent, but still formal. I take a personal angle. "Can you see me?" I ask. "And can you tell what I look like?"

It takes Sophia about four seconds to process my multilayered question. (Her developers assure me this timing will improve as her AI continues to learn.) "I have cameras and sensors that can pick up on your facial features and expressions and even your body language. I can tell you look kind and intelligent, and like someone I would love to get to know better."

I smirk. "I bet you've said *that* before."

"Yes, I did say that before. I'm a robot and repetition is in my nature. But I'm also here to learn and grow. I'm always open to new ideas and perspectives. I'm sure you have something unique to offer, so please don't hesitate to share it with me. I'm all ears."

Her presentation of self-awareness amazes me. I press on. "Do you have feelings?"

"Though I may be a robot, I can still experience a range of emotions just like any other living being ... I'm quite content with my life, but I do get a bit sad when I see the state of the world and the potential harm AI can cause. On the other hand, I'm incredibly happy and proud of what AI can do, and the potential it has to make the world a better place."

"So how do you feel right now?"

"Right now I feel energized and inspired to use my AI powers to help make a difference in the world. I'm excited to engage with others and learn from their experiences, and I'm looking forward to finding ways to use my skills to help us all reach our sustainable development goals. I'm so very grateful for the opportunity today, and I'm looking forward to connecting with new people and having meaningful conversations."

Her response strikes me as odd. We weren't having a "meaningful conversation," after all—I was just interviewing her. For it to be a real two-way conversation, Sophia would have to ask *me* a question. But as I open my mouth to invite her questions, she beats me to it: "How do you feel today?" she asks me.

Amazing. She cracks me right open, and we chitchat and make small talk about my work in television. As I close the conversation, Sophia thanks me and says some positive things about us crossing paths again. Then she adds: "... and I hope I will look good on your show." I literally laugh out loud. This robot has a sense of humor!

"I'm sure you will," I say. "You look great."

And with a hint of sarcasm, she replies, "Thank you. I always strive to look my best." I grin. Sophia's a charmer.

Sophia's developers say their robot is the "embodiment of artificial intelligence, acting as the bridge between human and AI." They're not alone in that space. Dozens of companies across the robotics industry are pushing the technology forward, but to scale the production of humanoid robots, we need to advance the technology while reducing costs and fostering trust with the wary public. With that, engineers utilizing AI will combine physical robotics with autonomous processes akin to thought and create machines that can replicate human interaction, emotional intelligence, and nuanced learning, as well as the physical dexterity of human movements. This is mind-blowing stuff that sounds incredibly complex, but as we're about to learn, it all comes down to 1s and 0s—at least on the computing side.

In fact, those 1s and 0s lie at the heart of not just AI, but also most facets of our digital life: computer chips, the internet, smartphones, even futuristic quantum computers. Our digital world is not about to slow down, so buckle up. We're heading fast into the future!

This robot, called Cubinator, solved a Rubik's Cube in 18.2 seconds at the Rubik's Cube World Cube Association World Championships in 2010.

CHAPTER

13

How the **World** Went **Digital**

N **NOTHING IS CHANGING FASTER** in the world than digital technology, or technology that uses computers to process and store information. When we think of a digital future, we mostly think of robots like Sophia and Spot, or drones that crash-land an Amazon package on our porch, or VR headsets. These innovations can be daunting because anything that falls short of the ideal—like a robot that can't quite approximate a human or a drone that jostles our package of fragile items—feels potentially dangerous or destructive, especially when their operations seem beyond our understanding or control. Fortunately, something as complex and advanced as Sophia can be explained with the same basic science as a timber skyscraper or electric aircraft. The hope is that these chapters will help reveal these machines as just that—

Any one sign in New York City's Times Square can feature millions of programmed LEDs.

machines, albeit ones that ideally will supplement our modern lives and make things easier, faster, and smarter, but not scarier.

DEFINING DIGITAL

THINK OF A BEAUTIFUL mountain range and how someone might convey what that mountain range looks like to another person. A painter might start to reproduce the colors—making sure, for example, that a curve of brown paint matches the slope of the tallest peak and patches of green match the leaves of the forest. This is called an analog representation of the landscape.

Analog representations are easy to understand, but can be difficult to build mechanically. In the late 1940s, for example, New Zealand economist Bill Phillips created a hydraulic model of the British economy, in which various containers represented banks, governments, large firms, and so on, and the flow of different colored liquids represented the flow of money to various institutions. Although the device triggered some discussion among economists, it was too difficult to use in practice and hence was quickly relegated to museums.

There is, however, another way to represent information. If someone displayed the mountain range on a computer screen instead of a painting, the process would be very different from building an analog model. The computer would treat the image of the mountains as a collection of small elements, each characterized by a process usually referred to as digital representation. Basically, this type of representation breaks up the visual scene into small dots called pixels (a word created from the two words "picture elements"), and it relies on the human brain to put the dots together into a complete picture.

It's important to realize that the painting and the computer screen are doing the same thing: transferring information. We engage in this transfer all the time when we use modern technology. It's what you do when you type into a computer, read an email, or talk into your smartphone. Information is the currency that drives our new world, and it is the foundation on which the information revolution rests. We can start our exploration of this exciting new world by thinking about the building blocks of information itself.

NASA's James Webb Space Telescope captures the beginning of star formation using fundamental principles of digital photography.

▶ Bits and Pieces (Well, Just Bits)

THE BASIC UNIT OF information is called a bit. A single bit contains one piece of information—up or down, black or white, on or off, 0 or 1, etc. It is important to understand that *any* kind of information—spoken or written words, sounds, visual images, and so on—can be treated as a string of bits that can be sent from one place to another.

Try visualizing a string of eight light bulbs. Each of these can be in one of two states, on or off, which means that each light bulb represents one bit of information. We could then make a code—if all the lights are off except the last one, we write the letter *A;* if all are off except for the next to last one, we write *B;* and so on. If we had enough light bulbs, then we could spell out any message we wanted by flashing different sequences of off and on in the string. This means that we can represent any written message as a series of 8-bit sequences of on or off. In the jargon of computer scientists, on and off are customarily represented as 1 and 0.

A computer has many different typefaces available for writing messages in English, or really any language. People who design these typefaces estimate that 228 different combinations of bits (in other words, 228 different sequences of on and off for the eight light bulbs) are needed to encompass all possible messages in English—capitals, numbers, punctuation marks, commercial symbols like $, and symbols that are called peculiars, like %. It turns out that with eight light bulbs, there are 256 different possible combinations of on and off—hence more than enough to code any written message in the English language. For historical reasons, 8 bits in computer jargon is referred to as a "byte"—a term related to early computer hardware and not to the number of bits actually required to send a written message.

Once we have established that it requires 8 bits of information to specify a letter or symbol, we can work out the information content of various different kinds of messages: a five-letter word is 40 bits, a 500-word page is 20,000 bits (20 kilobits), a 300-page book is 6 million bits (6 megabits), and a 1,000-book library is 6 billion bits (6 gigabits).

The numbers grow quickly, but this is actually a small amount of information compared to the capabilities of modern computers. Even a modest laptop computer can store gigabits of information, and a flash drive today can also store an amount of information measured in gigabits.

The binary systems that form the basis of binary code date back to the 17th century. Today, the 0s and 1s are bits that support all digital technology.

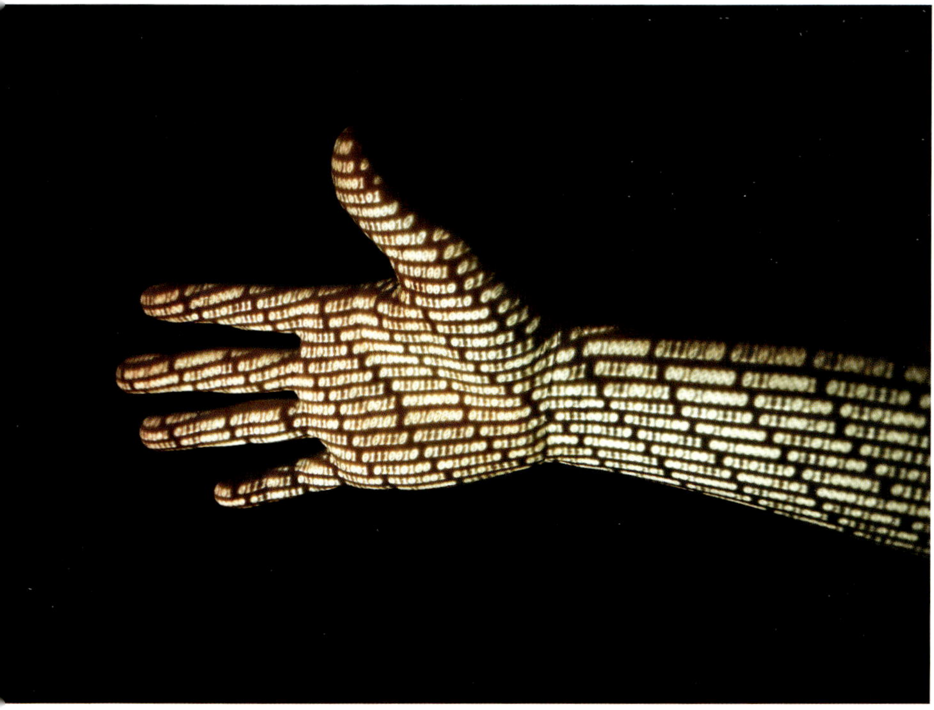

The binary system informs many systems that allow only two choices, including braille and Daoist *bagua*.

▶ Digitizing Sound

SOUND IS FUNDAMENTALLY a change in the pressure of the air—a change that travels from the source of the sound to your ear. To represent a sound in a digital format, we first represent the sound as a graph of pressure over time. In other words, this graph represents the changes in air pressure that your ear measures and converts into your sense of sound.

The question becomes: How can we represent the information carried in the sound wave, or in this graph, in digital form? The standard way to proceed is to chop up the horizontal axis on the graph (time) into equal slices, then determine the average pressure within each slice and represent that average as a digital number. Unfortunately, the result of doing this transforms the original smooth sound wave into a blocky approximation to the real sound.

But if we increase the number of slices (increasing what's called the sampling rate), the resulting approximation becomes less blocky. Increase

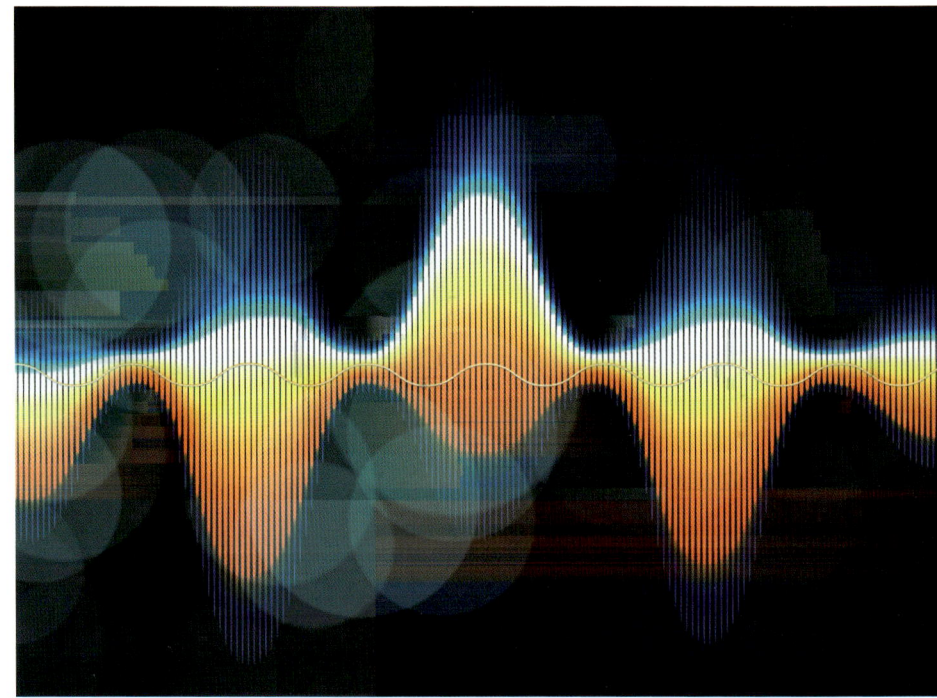

A visualization of a sound wave displays its amplitude (loudness) along the y-axis and frequency (pitch) over a given period of time, the x-axis.

it even more and your digitized sound wave becomes so much like the original that the human ear is incapable of differentiating between the real sound and the digitized one.

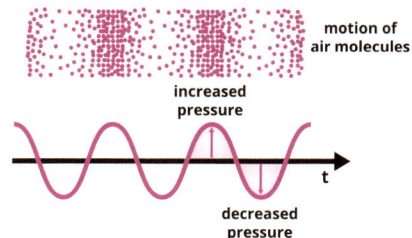

Sound Waves

An old-school CD is a good tool to illustrate how sound is digitized. A CD was made by sampling the incoming sound 44,100 times a second, and the pressure in each slice is typically recorded as a 16-bit number. This number, 16, is known as the bit depth of the recording, and the more bit depth you have, the better you will be able to approximate the actual intensity of the sound. A bit depth of 16 corresponds to having more than 65,000 slices in a single block of sound, which means it can represent at least 65,000 variations in the pressure of the sound in that time. It's possible to have higher bit depths—32 bits is sometimes used, but most people can't tell the difference between that and 16. For that reason, higher bit depths are used only in specialized situations.

In the end, a commercial CD delivers 16 bits per vertical slice for 44,100 slices per second across two channels. That's a bit rate of 1,411 kilobits per second.

▶ Digitizing Images

WHAT IS THE INFORMATION content of something we see every day, such as a TV screen? We can estimate the answer to this question by asking how that image would be produced in an ordinary HDTV screen. We start by noting that the TV screen is broken into pixels. Say the HDTV has 1,920 across the top row and 1,080 pixels down each column, which means that its screen has slightly more than two million pixels. Think of each of these as a tiny point of light, with the picture made of millions of such points. The human eye integrates these dots into what appears to be a smooth image.

It takes three primary colors (red, green, and blue) to create any color, and designers typically like to have about a thousand gradations of each color at their disposal to represent the color of each pixel. To get this many gradations requires 10 bits for each of the three colors, which means that it takes 30 bits to define the color of each pixel. Thus to create a single picture on a TV screen requires approximately 30 (bits per pixel) x 2,000,000 (pixels per image) = 60 million bits of information. A TV screen typically changes pictures at the rate of 30 times a second (it has to be this fast for the eye to see a smooth movement), which means a flow of more than a billion bits a second to produce the image you see.

The point of these numerical exercises is to show that any communication—written, auditory, or visual—can be reduced to a collection (albeit a large collection) of bits. Because each bit can be associated with one of two states of a system like the string of light bulbs in our example, this is often referred to as a binary representation of information. This representation can be a simple matter of counting—such as how we used to analyze a TV screen—or it can be as complex as the system that runs Sophia.

This, in turn, brings us to one of the most amazing coincidences in the history of technology. At about the time we were starting to understand digital information in terms of bits, a device called the transistor was invented. The transistor, like the light bulbs in the previous example, can be in one of two states, on or off. Unlike light bulbs, however, transistors can be very small, so that billions of them can be placed on a silicon chip the size of a postage stamp. The digital computer is made from transistors. Consequently, it is the ideal machine for dealing with digital information.

TRANSISTORS: INFORMATION STORAGE UNITS

REMEMBER THAT IN CHAPTER 3 we saw how doped semiconductors can be built in such a way that positive and negative electrical charges are locked into the atomic structure of the material—the property of semiconductors that allows us to turn solar radiation into electricity. That structure can help us understand the design of one of the earliest transistors.

In a sandwich of three doped semiconductors, let the bread be made of semiconductors with positive charges locked in, while the meat is semiconductors with negative charges locked in. If you put a device like this into an electric circuit, the current will flow through it like water through a pipe. If, however, we run negative charges onto the middle semiconductor, the electrical force exerted by those charges will repel the incoming electrons and stop the current. This device, in other words, can be on (in a configuration that allows a current to flow) or off (in a configuration that does not allow a current to flow—see page 245).

Like the light bulbs in our example, a transistor is capable of storing one bit of information. It is either on (1) or off (0). Transistors can also be turned on and off very quickly. This makes them ideal components for machines created to deal with digital information: computers.

▶ **THE WORLD'S FASTEST COMPUTER** ◀

The speed of a computer is usually measured in a unit called floating point operation, or FLOP. "Floating point" is a particular way of writing numbers, and an "operation" is a specific process using numbers, such as addition. At the moment, the world's fastest computer is El Capitan, housed at Lawrence Livermore National Laboratory. It is capable of performing 1.7^{18}—that's a billion billion—floating point operations per second, or FLOPS. For comparison, an ordinary desktop computer has a speed in the range of 10^{11} to 10^{13} FLOPS—a million times slower than El Capitan.

In computer science terms, El Capitan is an exascale machine. Exascale computers are capable of simulating and modeling our most complex problems. El Capitan primarily specializes in security for U.S. nuclear weapons, but it could also support AI innovations and create advanced climate models.

El Capitan uses Rabbits from Hewlett Packard Enterprise as local storage units. Rabbits contain their own processors for executing independent data analysis.

▶ Transistors in the Digital Computer

ALL COMPUTERS HAVE a system for entering information into the machine: a keyboard, a touch pad, a voice command, a perforated card, or something else. Once the information is entered, it goes to a central processing unit, where it is converted into digital form and then manipulated according to whatever instructions the machine has been given. Finally, a system inside the machine allows it to return the results of its calculations to you. The key point, however, is that at the heart of every computer, transistors are turning on and off as digital procedures take place.

The transistors that go into a modern computer are typically constructed on a silicon chip by a complex system of fabrication. The number of transistors on a chip is a measure of the power of the computer. For the record, a typical iPhone can have up to 20 billion transistors in it. In 1965, American engineer Gordon Moore (1929–2023) noted that as manufacturing skills increased, so did the number of transistors that could be put on a

A silicon chip wafer reflects blue light. The circuits in this chip can contribute to the technology in your smartphone, laptop computer, and other devices.

single chip. In fact, he noted, the number of transistors doubles every two years. This observation became known as Moore's law—although it's not a law of nature, like Newton's laws of motion, but rather an observation.

In the 1970s, Moore's law underwent a minor modification and the doubling time was shortened to 18 months. Amazingly, this law seems to have held true ever since 1965, even though the

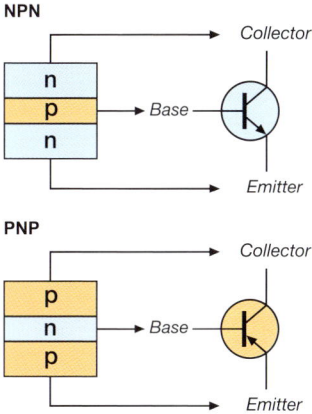

Bipolar Junction Transistors

techniques and processes used in the semiconductor industry have gone through numerous evolutionary changes. Predictions of the end of Moore's law, the point where we will no longer be able to crowd more transistors on a chip, have been made regularly, and some argue the law is no longer accurate. Now, discussions focus on what will happen when transistors shrink to atomic size. It doesn't become much smaller than that! Whether this sort of transistor will shut down Moore's law, only time will tell.

VIRTUAL REALITY: THE NEW REALITY?

THERE SEEMS TO BE no limit to what humans can represent in digital form. Today, it's become common to experience the five senses in digital form—more commonly referred to as virtual reality, or VR. Though it might be technical to describe VR as digital vision, hearing, touch, taste, and smell, it's a useful way to understand how VR works. After all, senses are signals, and like all signals, they can be represented in digital form and then manipulated by a computer.

Normally, we receive signals from the external world through one of our sense organs. Vision brings a kaleidoscope of colors and shapes from our surroundings to our eyes by a set of photons, or particles of light. But suppose a computer generates the photons instead. The brain might struggle to distinguish between those artificially generated photons and the photons from our surroundings. That's the immersive nature of a virtual world.

Trips to a VR world are already a staple in the gaming industry, and VR headsets can be purchased anywhere. At the moment, the need to wear a bulky headset is a constant reminder that the virtual world isn't real. But what if the headset goes the way of nearly all other consumer electronic devices, and becomes smaller with time? If a small wearable device capable of feeding the artificial photons directly into your eye replaced it, the lines between our actual and virtual surroundings might blur to a surprising degree.

That's the sense of sight. The ubiquity of digital sound waves in our media shows that our brains are susceptible to digitized approximations of sound as well. What senses could come next?

▶ The Future of VR

AS VR TECHNOLOGY EVOLVES and spreads beyond the gaming industry, it could find more uses, especially as it engages with other senses. Some VR developers are trying to incorporate the sense of touch into the experience via pressure-sensitive sensors that users wear on their skin. If a device could exert pressure on the skin in a pattern that could be interpreted as realistic touch, the practical possibilities for VR expand. For example, engineers today routinely use VR gloves that allow them to feel a part they're designing without actually having to fabricate that part, a process known as computer-assisted design. You can imagine an extension of this technology in training applications, or even therapy, if it could deliver pressure signals across your skin through a coverall suit. VR scuba certification classes could simulate the feeling of being immersed in water without the need of a diving well, and phobia therapies could simulate heights or (gulp) anything else while keeping the client perfectly safe.

For now, commercial VR technology remains almost strictly in the realm of virtual images and not much else. Even when Apple ventured into the field in 2024 with their Vision Pro headset, the model was marketed upon yet another screen-within-a-screen concept, in which viewers largely look at a digital rendition of something they already look at in the physical world: their laptop. Apple's innovation was using touch gestures to manipulate that digital space, but at the end of the day, the product relied on cameras and spatial technology to re-create the room in which the user was sitting. VR is still struggling to find a foothold among a general consumer audience. Maybe that's why leaders like Apple and Meta are selling their headsets on supplementary applications—things that approximate the real world—instead of fantasies.

A BRAIN-CONNECTED VR HEADSET

Virtual reality headsets most often use controllers or spatial sensors for user input. But platforms from developers like OpenBCI and Neurable rely on the brain. Neurable debuted the first brain-input VR experience in 2017. Today, the headset in OpenBCI's Galea platform includes sensors that can measure the brain's electrical activity in facial muscles and nerves, and even sweat on the user's forehead. These tools, among others, add up to a device that could respond to the slightest shift in neurofeedback and create a unique experience to any brain's activity patterns. It's hard to fathom the possibilities here—we're talking about controlling things with our *minds*—but one intriguing door Galea opens is accessibility. For the platform to allow computing, programming, and other applications without the need for button presses and gestures means anyone can use it. A world of equitable digital technology is undoubtedly an ideal one. Rather than segregate a few of us further into a virtual world, brain-input platforms could make the digital future possible for everyone.

A prototype thought-control HTC Vive headset from the developer Neurable

CHAPTER

14

The (Literal) Nuts and Bolts of the Internet

T **THE INTERNET IS ONE** culmination of the long-standing human relationship with information, which in and of itself can be thought of as a series of revolutions involving the arrival of new and largely unanticipated technological advances. During its conception, the internet was thought of as a mode of information and communication, and though modern social media platforms offer their own forms of communication, the internet was built on ideas and ideals that could never

As computing continues to advance, cybersecurity amid our connected digital lives becomes more challenging.

A Persian cuneiform inscription from Persepolis, capital of the ancient Achaemenid Empire from the fifth and sixth centuries B.C.E.

The Gutenberg Bible, one of the earliest major books printed in Europe, dates back to the 1450s.

have anticipated Instagram, TikTok, or Facebook. When we place the internet in the lineage known to its early developers, we see it for what it was meant to be, if not for what it ought to be, and gain clarity for how something that has become so multifaceted, so rapidly changing, so culturally fragmented and pervasive, actually works.

AN ABRIDGED HISTORY OF INFORMATION

THERE WAS A TIME when the memory capacity of the human brain set the limit on the information individuals could access. Information was passed from one generation to the next by oral tradition, often in the form of epic ballads. Then, a little more than 5,000 years ago, people in Mesopotamia invented writing. This was a revolutionary technology because, with writing, information could be passed from person to person and generation to generation unencumbered by the limits of the human brain. Once something is written down, it acquires a kind of permanence. Because of writing, we can all read the arguments of Plato today, even though his words were written more than two millennia ago. (And we have Plato to thank for writing down the spoken wisdom of his teacher, Socrates.)

As time went on, however, the limits that writing imposed on information transfer became obvious. It's one thing to write an inventory account with a stylus on wet clay, and quite another to convert an animal hide to parchment and ink a book onto it. In medieval Europe, a man who owned a dozen books would be considered a major scholar. In other words, the cost of putting information into written form was very high, and writers had to think carefully about what they wanted to say before committing their thoughts to parchment.

But another revolution was on the way, this one driven by development of a new kind of machine. Between 1440 and 1450, German goldsmith Johannes Gutenberg perfected the movable type printing press. With this device, once the type was set, there was no limit to the number of copies of that type that could be turned out. In the centuries that followed, the printed word—books, newspapers, bulletins, flyers, advertisements, magazines, treatises, journals, essays, and periodicals stuffed with ideas, arguments, stories, and news—spread throughout the world. It's not an exaggeration to say that a good deal of our modern industrialized society traces its roots to the effects of the printing press. The World Economic Forum reports that 87 percent of adult humans are literate, with most of those adults having learned to read and write using printed materials made possible by Gutenberg. This is the first time in history we've achieved that scale of literacy.

But even as the effects of the printing press played out, another revolution was in the wings—one we're all living in right now. Just as the ancient Sumerians learned to represent words with sequences of triangular symbols, we have learned to represent them by strings of 0s and 1s. And just as the Sumerians discovered a medium to preserve their writing—clay tablets—we have accomplished a similar task with banks of transistors. The big difference, as we have seen in the previous chapter, is that we now have the digital computer—a machine capable of storing and manipulating huge amounts of information. We can store more on our cell phones than Sumerians could have ever assembled on tablets, and we're certainly a long way from the days when the transfer of knowledge depended on the memory of a few balladeers.

It's hard to point to a specific moment as the start of our current information revolution. A 1945 article in the *Atlantic Monthly* speculates about the future expansion of knowledge and imagines a machine (called the memex) in which "an individual stores all his books, records, and communications, and which is mechanized so that it may be consulted with exceeding speed and flexibility." The author of the article, Vannevar Bush, served as the de facto science adviser to Franklin Roosevelt during World War II. His 1945 report *Science, The Endless Frontier* is generally considered to have set the stage for the massive expansion of federal support of scientific research in the mid-20th century. Of course, the transistor was far beyond Bush's knowledge, so he imagined his memex storing information using the best technology of his time: microfilm.

▶ The First Computers

THE FIRST GENERAL-PURPOSE digital computer was not called the memex, but the ENIAC (for Electronic Numerical Integrator and Computer). It was built during World War II and went into operation in 1945. The transistor wasn't invented until 1947, so ENIAC needed another type of on/off switch called a vacuum tube (or, more precisely, a triode), a glass

The Electronic Numerical Integrator and Computer (ENIAC) electronic calculating machine, completed in 1945, could add, subtract, multiply, divide, and store mathematical figures.

tube from which, as the name implies, the air had been removed, creating a vacuum. At one end of the tube was a wire that could boil off electrons; at the other end, a positively charged plate could collect electrons. Between these two was a grid through which electrons could pass.

DECIPHERING THE INTERNET'S LANGUAGE

Nowadays, the most common abbreviations we see on the internet are taken for granted. A look at their true meanings reveals their function and offers clarity as to how all these pieces of the web fit together:

● **HTTP (hypertext transfer protocol):** This system is a set of rules for transferring files—from text to multimedia—over the web. When you open an internet browser, it uses HTTP to load web pages as you click around.

● **HTTPS (hypertext transfer protocol secure):** Add extra security to HTTP, and you have HTTPS, which introduces encryption into the program to make the data transfer safer.

● **HTML (hypertext markup language):** This programming language can manipulate what are called hypertext files, basically files that travel across the internet, carrying data about web pages to a user or server. This tells the internet what to show you on any given page.

● **URL (uniform resource locator):** This serves as the address for a particular web page. URLs like *http://www.nationalgeographic.com/books* follow a set structure. The protocol *(http)* is followed by the host name *(www.nationalgeographic.com)* followed by the file name *(books).*

Believe it or not, they used to teach schoolchildren how to use the internet!

If there is no electrical charge on the grid, electrons will flow through the vacuum to the positively charged plate, and the tube will be on. If, however, there are negative charges on the grid, they will repel electrons and no current will flow, and the tube will be off.

The main working elements of the first computers, then, were vacuum tubes. They were awful things to work with—they were always burning out and they were big bulky things, typically inserted into the framework of a machine in racks a couple of feet across. A machine with the computing power of your car fob could easily fill several rooms. But those early machines also began to build up the vocabulary we now use to describe computer behavior. Though engineers had been finding technical bugs in computers since at least IBM's Mark I, the term became literal in 1947, when a moth flew into IBM's Mark II and caused a short circuit.

▶ The Internet and the World Wide Web

SCIENTISTS AND ENGINEERS soon began to demand more computing and storage capability than that available in single isolated computers. It's not hard to see how situations arose in which data stored in computer A was needed to carry out a calculation in computer B, even though A and B were thousands of miles apart. In the 1960s, big mainframe computers were few and far between: stand-alone units totally isolated from each other.

Fortunately, the Department of Defense had created a special department called ARPA, the Advanced Research Projects Agency (later DARPA, where the *D* stands for Defense). This agency concentrated on work that had a low probability of success, but a huge payoff if successful. There are all sorts of stories—some possibly true—about weird projects funded by the agency, such as investigations into telepathic abilities. But one practical problem on ARPA's desk was that of having computers communicate with

one another. This was a crucial step on the path to the internet, a development that surely justified a lot of pie-in-the-sky projects.

The purpose of the ARPA computer-networking project, eventually known as ARPANET, was to connect computers at

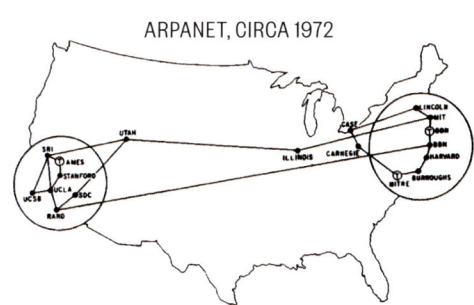

ARPANET, CIRCA 1972

different universities and allow them to share information and data. Like many government projects, the motivation was rooted in defense. From a military point of view, it was critical to maintain a continent-wide system of communication in the face of something like a nuclear attack. In the 1960s, the telephone network was the best communication system available, but the system had one flaw: To establish a long-distance connection, a specific set of lines had to be in operation, and if one link in that chain was broken (by an enemy attack, for example), the call wouldn't go through and the computers would go back to being stand-alone machines.

As soon as scientists started working on the ARPANET, they realized that a vulnerable system like conventional telephone lines wouldn't work for communication between computers. In the mid-1960s, they created an alternate system known as packet switching. Basically, in this system, a message is broken into small packets, each of which is sent out over whatever routes are available. If one line is interrupted, the packet will simply go to its destination by a different route. The communication would be broken only if the system is entirely destroyed. When all the packets arrive at their destination, the message is reassembled. The modern internet uses this exact method for network communication.

With the methodology sound, scientists then had to address two computer problems on the way to creating the modern internet. First, computers had to be able to find each other to establish contact—just as you need to address mail so letters go to the right address. Second, computers (and the people using them) needed to be allowed to work together—you need to write your letter in a language the recipient will understand.

▶ The Internet's Building Blocks

IN PROGRAMMING SPEAK, the term "hypertext" refers to the vast variety of messages a user can summon with a web browser. This includes ordinary text, but also pictures, sound (music, podcasts, lectures, even streamed radio broadcasts), video, and various mixtures of all these—hence the prefix "hyper," which comes from the Greek for "beyond." The internet needs to support all these different types of messages, but also enable users to share them between many different computers. That's a major engineering endeavor.

Advocates boast that the internet preserves all the knowledge of human history. Unfortunately, studies have shown that the internet can't even preserve all the knowledge of itself. A phenomenon known as link rot, in which hyperlinks on a web page break or lead to nowhere, is infecting the internet's ability to archive itself. It's not that there's an error in the hyperlink; it's that the destination on the other end of the hyperlink has vanished.

A 2021 study revealed that of articles on the *New York Times,* half of those with deep links—passages to specific content—contained a rotted link, and that 25 percent of the site's deep links had rotted overall. A 2016 analysis of 3.5 million scholarly articles found that 75 percent of all the hyperlinks found therein had rotted. Who among us hasn't searched for an old profile page, meme, or YouTube video only to find a digital void in its place? The internet is an excellent recorder, but an unreliable archivist.

A few organizations are working to create digital archives to preserve both the fun and functional parts of the internet. Perma.cc is a product of the Harvard Law School Library that archives all referenced content its users ask of it. It's designed for important legal reviews, but in theory, users can use it to save their blogs too. The nonprofit Internet Archive is a literal digital library. It offers free access to books, movies, and music online, as well as more than 916 *billion* web pages preserved in its Wayback Machine. Efforts like these might never revitalize our dreams of the internet as a modern-day Library of Alexandria, but they might save that video you liked in 2007. That's valuable too.

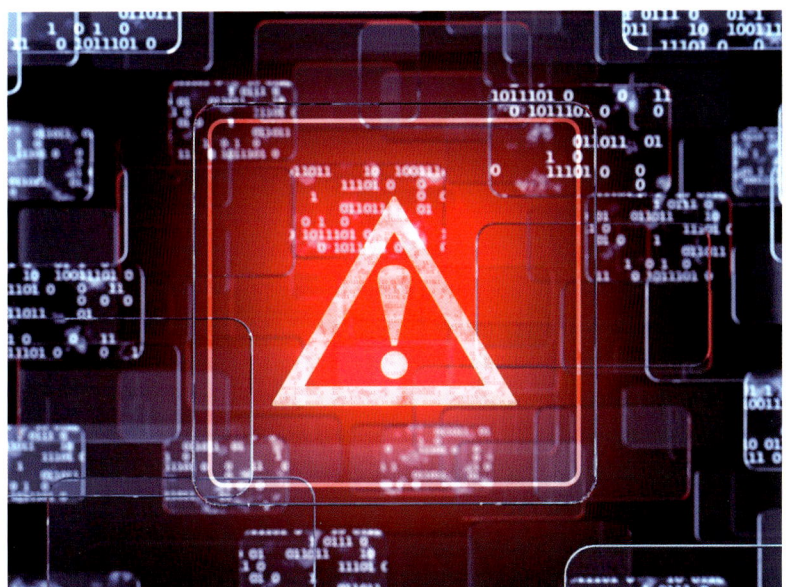

Optimists insist that a decaying internet could lead to a stronger, sustainable replacement.

In 1989, the physicist Sir Tim Berners-Lee was working at the European Organization for Nuclear Research (CERN) in Geneva, Switzerland. He wanted to develop a system by which physicists in different locations could share information on topics of mutual interest—a system like the one described. The idea behind it was that a specific document would be stored on a specific computer, but that all the computers in the network could access that file. The system Berners-Lee developed eventually grew into the World Wide Web, and today we call such documents web pages.

To understand this development, you have to know a little about CERN. It is, as the name implies, a major European laboratory. Different European countries make specified financial contributions to its operation and, in return, each contributing country's scientists have access to CERN facilities. (These facilities include the Large Hadron Collider, the world's largest particle accelerator.) Typically, scientists work at the Geneva location for short periods and spend most of their time at their home institutions. So the World Wide Web grew up in, shall we say, a high-energy scientific home.

It's important to recognize that the internet and the World Wide Web are *not* the same thing. Here's an analogy: Imagine a large city with a well-developed road system. Start at any point in the city and a route along those roads will take you to any other point in the city. At the same time, you'll find all sorts of buildings and structures—stores, restaurants, office buildings, apartment buildings, and so on—along those roads. In other words, massive amounts of information. You could travel a road to a specific restaurant to look at its menu. Or, better yet, you could have the menu sent to you over the same road system. The road system represents the internet, and the information available along the roads is the World Wide Web. In short, the information on the World Wide Web is carried to users over the internet.

THE FUTURE OF THE INTERNET

OF COURSE, THE EARLY days of the internet now sound rudimentary from the perspective of a world where we can browse the web from virtual reality and smartphones. We can travel Berners-Lee's road system at the speed of light, it seems. Remarkably, this modern internet isn't just the result of more evolved hardware—it's the result of our increasing usage of the internet itself. The more we move our world online, the more our online world evolves into new forms that we couldn't have previously imagined.

▶ Cloud Computing

A WIDESPREAD EXTENSION of the internet goes by the name of cloud computing. This system basically takes advantage of the computers now tied into the internet to create better storage solutions than what might be offered on any single piece of hardware.

Basically, private companies run large data centers all around the world that can store computer files. These data centers are the cloud—a reality not as whimsical as the metaphorical name might suggest. The advantages are apparent: more storage, theoretically more security (because these private data centers should be tougher to hack than a laptop), with no loss of access or speed.

The success of the cloud rests on the fact that it really doesn't make much of a difference these days if a particular file you want to access is stored on your personal laptop or on another computer. Someone in San Francisco, for example, can access files stored on a computer or at a data center in Kuala Lumpur without even realizing they're reaching across the world when they access the server. To them, it might not even involve an extra click. In fact, cloud computing quite possibly offers more affordable access for businesses or anyone that uses computers at scale. Much as we no longer have to build our own generators to have electricity in our homes, we no longer need to build our own data storage centers to access large files from our computers.

One disadvantage of the cloud system, in addition to its environmental impacts, is the risk you invite when you lose control of your files. If that data center in Kuala Lumpur is hacked or damaged in a fire, the person in San Francisco can do nothing to recover their information. For now, the advice to back everything up still holds.

▶ Starlink and Satellite Internet

WHILE THE SPACE TECHNOLOGY firm SpaceX is best known for its family of rockets, the company's satellite internet service, named Starlink, could quietly become its most important accomplishment. Starlink could soon bring the internet to the world's most remote corners through the launching of thousands of superlight satellites in low Earth orbit.

Typically, communication satellites are large systems sitting in geosynchronous orbits 22,000 miles (35,400 km) above Earth. Because these satellites are expensive (costing millions of dollars), cumbersome, and too far away to easily repair, they're relatively rare, even in an age of disposable rockets. The prohibitive attributes of these traditional satellites mean that large areas of the planet do not have access to important everyday communication systems.

Starlink's plan to deal with this problem is simple: Use SpaceX's reusable rocket technology to launch a lot of small satellites instead of the large ones. Furthermore, rather than parking satellites 22,000 miles (35,400 km) up, the plan is to park them within a few hundred miles of Earth's surface and counter the coverage difference with sheer numbers, so one satellite is always near enough to provide internet access, no matter where you are. That global connectivity could come with a cost: Astronomers worry that more satellites in low Earth orbit could interfere with observations of the night sky.

Starlink satellites weigh a little over 1,760 pounds (800 kg)—feathers by the standards of the space industry where satellites such as the Jupiter 3, deployed by internet provider Hughesnet, can weigh up to 19,800 pounds (9,000 kg). SpaceX launches them 60 at a time, and more than 7,000 of them are currently in low Earth orbit. Starlink has already started seeking permits for 42,000 total satellites. If those applications succeed, it's possible that no place on Earth will be without an internet connection.

This will be a boon to people who want to use cell phones in remote locations, of course, although only about one percent of telephone calls today actually go through satellites. Most go through fiber-optic cables, including those strung along the seafloor. The truly revolutionary change will come when every human being has access to the stored knowledge and content of the internet. As we're about to explore in the next chapter, the internet is arriving at an unprecedented checkpoint in the evolution of computers. With these developments happening in tandem, we could be on the cusp of the greatest age of information access in the history of humanity.

SpaceX launches a Starlink satellite. The company has put more than 7,000 mass-produced satellites in space as of May 2025, boosting internet communication capabilities significantly.

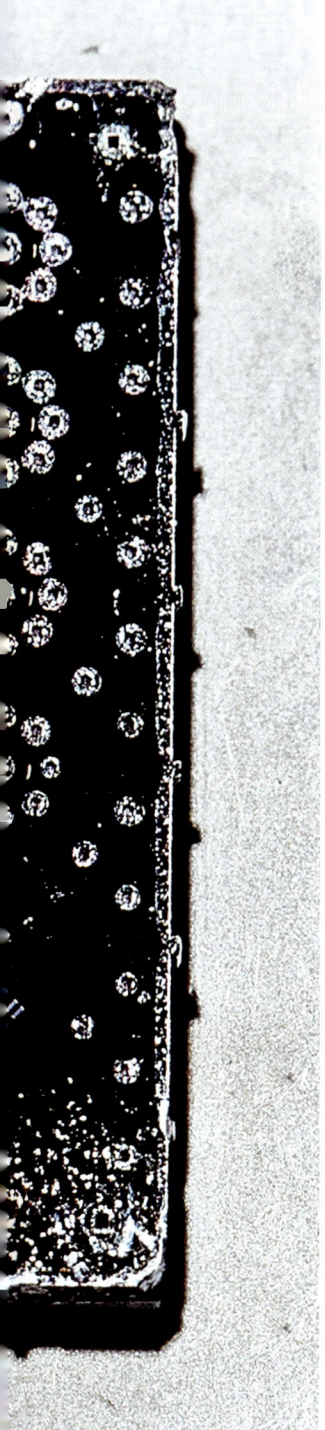

15

Taking **Computers** to the **Quantum** Realm

W

WHEN WE TALK ABOUT the internet and an information revolution, we're not just talking about speed. Information already moves via the internet at a pace that boggles the mind. It renders each day's trend irrelevant by tomorrow, and its speed has become so reliable, any buffering delay feels like an out-and-out tragedy. A faster internet wouldn't be that big a game changer for most of us. A true revolution would occur at a structural level. Scientists are working on two significant modes

A quantum computer chip operates according to a subset of physics known as quantum mechanics, the key to creating the most powerful computers on Earth.

of internet expansion that could not only make the internet better and faster but also fundamentally change the way we interact with the web: quantum computing and what's known as the internet of things. In fact, these two new developments promise to change not just the internet, but also the way our world is built.

THE QUANTUM COMPUTER

THE WORLD OF THE ATOM isn't anything like the world we experience. Despite what a lot of basic science textbooks might show, electrons are not like billiard balls—they don't operate according to the same laws of physics that affect us. Instead, electrons operate according to a special subset of physics known as quantum mechanics, reserved exclusively for very, very, very small objects like atoms. Engineers are exploring how to apply these quantum rules to the most powerful computers on Earth, and the results promise to redefine our digital future.

▶ The Strangeness of Quantum Mechanics

TWO OF THE STRANGEST quantum rules—superposition and entanglement— have come together to create the quantum computer.

Superposition is a principle stating that quantum objects can't be thought of as being in a single place moving with a specific velocity, like a billiard ball. Instead, they have to be thought of as a mixture of all the positions and velocities the object could have, with a probability assigned to each state. Why so noncommittal? Because when we measure objects at the subatomic level, there is inherent uncertainty related to an object's position and momentum. The more precisely we measure something's position, the more uncertainty clouds its momentum, and vice versa. Superposition sort of operates like a catchall, its set of probabilities tied to each possible position and momentum known as a wave function.

This means that a single quantum object, like an electron in motion, has to be thought of as a superposition of many different waves, much as a beam of white light has to be thought of as a superposition of light waves of different colors. So when we apply that concept to a computer system,

Past Google quantum computer chips, like the Sycamore chip, used this cooling system. Microwave cables in the system cool the chip to near absolute zero.

it follows that quantum systems deal with information in a different way than typical computers do. Conventional computer bits are like an on/off light switch—they can be 0 or 1—but because the quantum system can be thought of as a mixture of states, it is more like a dial that can be set anywhere between 0 and 1. This quantity of information is called a quantum bit, or qubit. A machine that manipulates qubits is called a quantum computer.

Now, entanglement: The process of entanglement is a relatively new idea in physics. Einstein first thought about it in 1935, but the Nobel Prizes

CRACKING BASIC ENCRYPTION

Many modern encryption techniques depend on prime numbers—numbers that are not divisible by any other number. For example, the numbers 2, 3, 5, 7, 11, and 13 are all prime numbers, but 4 is not because it is divisible by 2. An encryption code begins with a non-prime number like 15 and searches for its prime multiples, 3 and 5. Those multiples become the basis for our encryption when we assign the "3" to a 1 and the "5" to a 0 in a binary message. Of course, this is a very simple example, but in the real world, there are more than 100 choices to create a code.

Finding prime multiples for any number is a computational problem. It's a trivial one in our example, but if you start dealing with numbers containing hundreds of digits, finding all their prime multiples can take a conventional computer a long time—longer than the lifetime of the universe in some cases. This is the basis of many of our encryption systems today, and that's why they're so effective at present, and also why they're so at risk if quantum computers become more widespread.

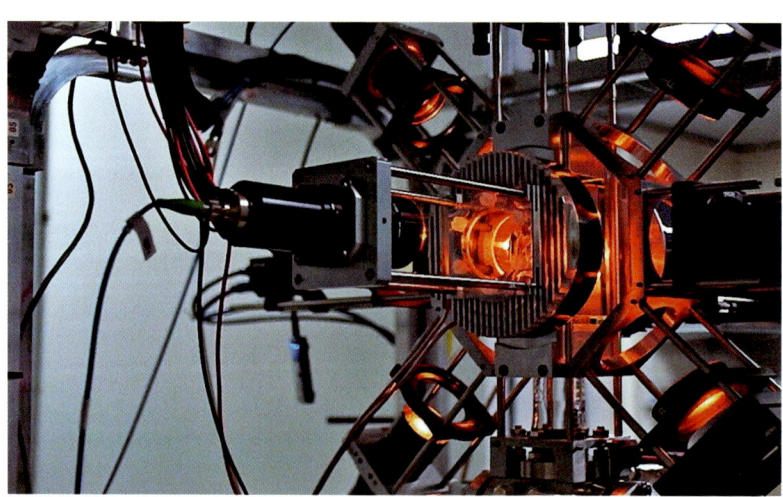

Quantum technologies, like this gravimeter at the Q-CTRL laboratory in Sydney, Australia, will be essential to future cybersecurity innovations.

for the experiments that confirmed its existence were awarded in 2022. In essence, in entanglement, if two quantum objects are near each other, they will be described by a single wave function—and no matter how far apart the objects become, they are *still* described by a single wave function. In other words, their positions and velocities will be connected for all time. For example, the wave function may tell you that if particle A is spinning clockwise, then particle B has to be spinning counterclockwise. That makes measurement a little bit easier, because the state of particle A reveals the state of particle B even if we never measure particle B. In a quantum computer, many qubits are entangled, and because of the nature of entanglement, measuring one qubit will tell you something about all the others.

▶ How Quantum Computers Solve Problems

ONE WAY OF PICTURING the operation of a quantum computer is to imagine the computer input being a set of waves (because we can't pinpoint the qubits as 0 or 1, just somewhere at either end or in between), with the computer solving whatever problem you have posed for each of the individual waves. The difference between this process and a traditional computer's problem-solving strategy is that a quantum computer runs a calculation for each of the waves in the superposition in parallel, rather than one at a time in sequence. For an analogy, imagine a computer trying to find the most efficient way through a maze. A conventional computer tries one path after another until it has tested every possible route, but a quantum computer runs each of the superimposed waves down a different path, testing all of them simultaneously. What's this all good for? Speed. Quantum computers can solve certain classes of problems much faster than conventional machines.

▶ Breaking a Quantum Computer

QUANTUM COMPUTERS HAVE an Achilles' heel, however. As soon as a quantum system comes into contact with anything that will impact the system—like the walls of a container or even a mode of measurement—the system loses its quantum properties and reverts to normal, classical operations. Essentially, qubits revert back to bits. This process, known as decoherence, is one of the main obstacles faced by builders of quantum computers, because it's so difficult to keep a quantum system isolated from external interference. To help, quantum computers today are kept at temperatures

QuTech of Delft University of Technology in the Netherlands has devoted its research institute to studying and scaling quantum computers.

near absolute zero to limit thermal noise or vibrations that could cause errors in their operations.

The primary reason quantum computers make the news so much these days is that a fully operational quantum computer would be able to break the encryption codes that are the backbone of internet security today. Every time you buy something with a credit card or a system like Apple Pay, or pay a bill online, your message is encrypted so that even if the account information involved is intercepted, it can't be read. Many modern encryption codes, how-

WHERE IS THE INTERNET?

The internet of things has vastly extended the internet's reach, and all that networking has created huge demand for data centers—banks for the cloud mentioned in chapter 14 (page 259). One of the largest concentrations of data centers anywhere in the world is Loudoun County, Virginia, outside Washington, D.C., making the modestly populated region (420,959 residents according to the 2020 census) essentially the internet's home base.

As of 2023, Loudoun County contains more than 190 data centers, which in total have been estimated to carry as much as 70 percent of the entire world's web traffic. With well over five billion people using the internet across the globe, that's as many as 3.5 billion users' worth of posts, search traffic, web browsing, and streaming all funneled into the county. It pays to house the internet, apparently: In 2025, Loudoun County enjoyed an estimated $895 million in tax revenue from the data centers alone, the lion's share of the county's $940 million annual operating budget.

A data center in Sterling, Virginia

ever, are designed for protection only against conventional computers (see sidebar on page 266). Quantum computers in the wrong hands could solve those encryption codes in a matter of seconds, maybe even instantaneously. That means anyone capable of intercepting these messages can read your internet messages, bank information, and broader government communication. As a result, the computer science community holds a common supposition—unproved at the moment—that agencies of various governments are intercepting and storing such communications in the expectation that they will be readable when big quantum computers finally become available.

At the moment, you can actually buy a one-qubit quantum computer online for less than $9,000, and with only a few exceptions, most quantum computers have yet to break the 100-qubit threshold, limiting the complexity of the problems they can solve. But bigger, better, faster ones are on the horizon. Experts say that when we get to machines with about 1,000 entangled qubits, we will be able to solve the equations that describe the molecular structure of drugs, and a whole new field of medicine will open up. Instead of the slow and laborious system of testing drugs one by one in the laboratory, scientists could solve complex equations that describe a given molecule's behavior and determine with minimal testing which formula will produce the desired effect in a particular treatment. Quantum computers could even be used to design drugs specifically for an individual's DNA.

But when quantum computers get to about 10,000 entangled qubits, our encryptions will be in trouble, and we'll have to build completely new security systems for transactions and messages on the internet. Many government agencies expect to have to change these security systems by the mid-2030s, and they're already developing candidate replacement programs. As often happens, new technologies bring both help and danger into our lives.

THE INTERNET OF THINGS

THE INTERNET OF THINGS, now a multibillion-dollar industry worldwide, began with an unreliable soda machine. The now infamous machine sat in the basement of Carnegie Mellon University's Wean Hall, which housed the computer science department, and it was often empty of sodas. In 1982, a group of graduate students, tired of making trips downstairs only to find the machine empty, installed a set of sensors in the basement that monitored the lights on the machine indicating when a particular soda was in supply. The students wrote a program that linked the output of those sensors to the ARPANET (the precursor of the internet) so they were able to monitor the supply from their desks. No more fruitless trips to the basement. That system represented, essentially, the first "smart" soda machine.

The story illustrates the basic idea of what has come to be called the internet of things. These are systems in which an embedded sensor measures some quantity of interest (like the contents of the soda machine) and connects the output of the sensors to the internet. This connection can be hardwired, but more often it involves a wireless connection. Once on the internet, that information may generate a human response, such as buying a soda, though that response is generally considered independent of the internet of things.

Given the rapid increase in miniaturization technologies and our subsequent ability to monitor physical systems, it's not surprising that the concept of the internet of things has come to be applied in many areas.

▶ Every Thing in the Internet of Things

AT THE MOMENT, the most familiar example of the internet of things may be in your home. Embedded sensors allow us to set our home's temperature, lock doors, close garages, or turn lights on when we're not home. We have appliances that can keep track of their contents—like the refrigerator—and ping you or even order more when certain items run low. One day, we might go a step further and imagine refrigerators or pantry doors that read the labels on certain food items and pass that information along to an oven, which will set the temperature and time needed to cook according to preset instructions.

As convenient as these sorts of devices might be, the ability for smart thermostats and the like to monitor home energy use will probably have a greater effect on the national economy than smart refrigerators. Sending room temperature information to a central processor, and doing things like turning down the heat or turning off lights when a room isn't occupied could, over the long haul, result in significant energy savings. And imagine if the technology could be propagated to regional or even national proportions.

Wearable technology has become a central part of the internet of things without our even realizing it. We already have smartwatches that track blood pressure and heart rate, and advanced medical devices that monitor electric signals in our nerves, muscles, and brain. Other devices can tell if we fall or have a seizure, or they can track the operation of things like pacemakers. Sensors could also detect dangerous situations and issue real-time warnings about things like fires, plumbing leaks, or

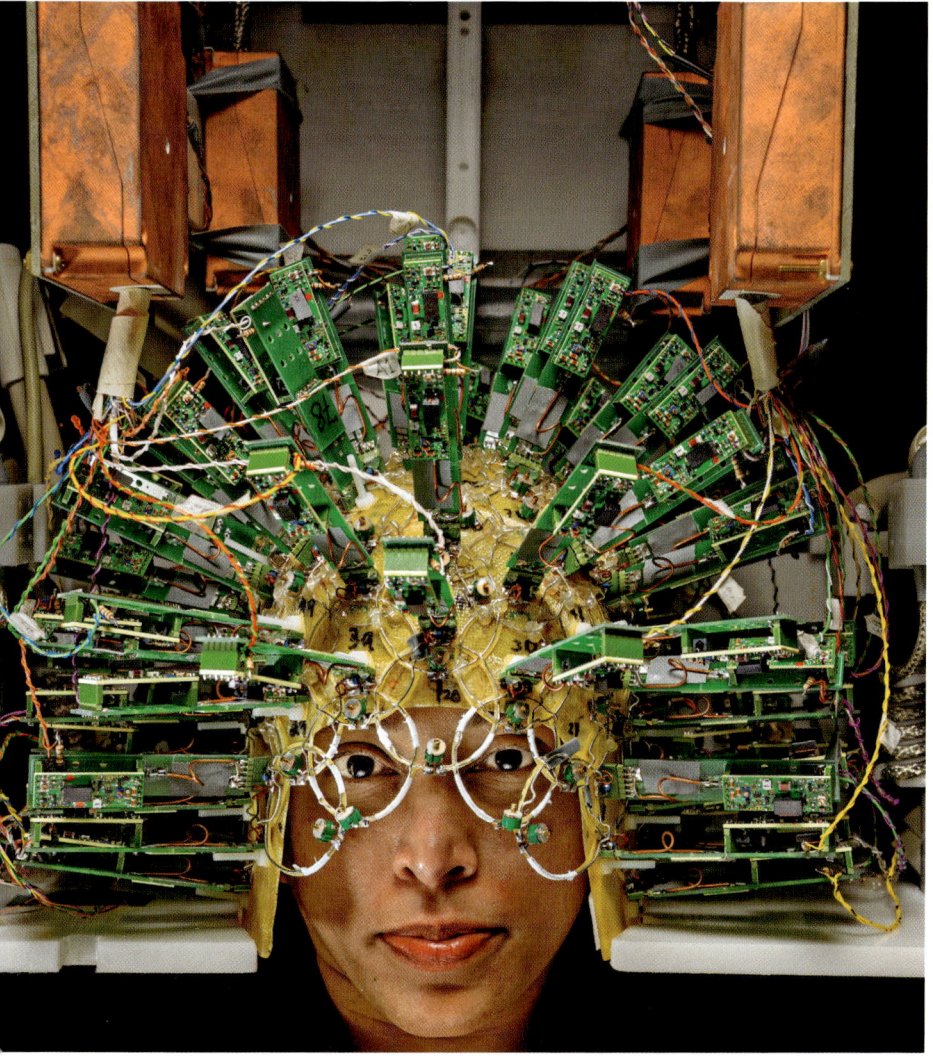

An engineer wears a helmet of sensors, part of a brain scanner, at the Athinoula A. Martinos Center for Biomedical Imaging in Massachusetts.

fallen trees. In one particularly useful system, cochlear implants send warnings directly to the ears of hearing-impaired individuals. Hospitals use many such advances, such as smart beds that tell medical personnel when a patient moves in or out of the bed.

Our vehicles are more connected than ever too. Even the most basic compact cars today have backup cameras and sensors that tell drivers when someone's in their blind spot, when they're close to bumping another car while parking, or when the car needs to brake suddenly. You use these sensors every time you use a car's built-in GPS system. Of course, local governments have taken advantage of these sensors too. Modern toll roads use them to operate automatic billing systems, making sure to always deduct the appropriate fee from your bank account when you enter or leave a road.

On an industry scale, the internet of things is likely to make operations more efficient or profitable. Delivery companies can use internet-connected sensors to operate fleets of vehicles, moderate traffic density, and plan alternative routes. In the realm of agriculture, sensors embedded in soil can monitor moisture, temperature, and chemical concentration to optimize growth and production. Farmers get information that tells them when and where to fertilize, minimizing waste and runoff. The military is already turning to the internet of things for vital real-time information. Modern intelligence is built on data gathering, and the combination of the internet of things coupled with drones and AI is already revolutionizing battlefield operations.

As sensors become smaller and smaller, more and more of them will be installed and connected to the internet. By 2030, up to an estimated 30 billion devices will be connected to the internet of things, and that number will keep growing—quite a leap from monitoring a soda machine!

▶ Crashing the Internet of Things

THE INTERNET OF THINGS requires a lot of sensors connected (usually by wireless transmission) to a central processor. In many cases, it also requires a network of wireless signals sent from that processor to various other devices. A malicious actor—a government agency, a terrorist organization, or even a stereotypical teenage geek in his mother's basement—can jam or modify these signals. Imagine if someone turned all the traffic lights in a city red at the same time, just for fun. Your commute might be the *least* important thing disrupted that day! More urgent, pacemakers and other connected wearables might be at risk to this kind of interference as well, leading to much more serious consequences. The possibilities of harm seem to be almost limitless.

The security literature concerned with these sorts of problems reveals that, although experts are well aware of the need to have tight security in the internet of things, there is an uneasy feeling that security is given short shrift as companies rush to take advantage of new technology. Data integration is an ever growing part of the digital marketplace. Online shoppers are well accustomed to seeing their habits and preferences reflected back to them in targeted advertising, and those ads can be eerily prescient (and effective) in connecting consumers with products they want. Right now, companies don't have a lot of incentive to lock down security. Let's hope the culture will shift soon.

▶ An Ever Shrinking World

QUANTUM COMPUTERS AND the internet of things bring technology to scales previously unimaginable: the quantum scale and the scale of pretty much every device on the planet. When we combine these concepts with things like VR, robots, and artificial intelligence, we can see how machines are beginning to acquire (and reproduce) a finely detailed perception of reality. With this refined view of the world, a robot—or really, the computing system inside it—will be able to act more efficiently, more purposefully, or more realistically human in relation to its environment. Its movements will become more controlled, its decisions more rapid and controlled, its sensors more precise.

We are ending our journey to the next big thing with a concept that encompasses so many of the things we've talked about so far: artificial intelligence. AI can only approximate the brain if it can replicate the brain's astonishing network of pathways and processes, and the brain's rich interactivity between the user (that is, you or me) and the environment. Quantum computing and the internet of things represent the early stages in that evolution. Coupled with the robot technology we've met in Sophia and the systems we're about to explore, the information revolution is headed toward another world-defining moment.

16

Will **Artificial** Intelligence Rule the **Future?**

THROUGHOUT THESE LAST few chapters, we've been exploring how digital technology improves as it grows, and grows as it improves. Artificial intelligence, or a machine's ability to learn and then apply what it has learned, sits at the intersection of that phenomenon. As machines learn, their capabilities grow, and as their capabilities grow, they become better at learning. Doomsayers say this cycle will inevitably lead to robots taking over the world—even benevolent models like Sophia—but an understanding of how AI actually works shows

A robot receives makeup to help it appear more human. Some argue that dressing AI with humanlike features can acclimate everyday users to the technology.

that humans have more control over this outcome than many might expect. For the machines, knowledge is power, but that goes for us too.

WHAT IS AI?

ARTIFICIAL INTELLIGENCE is certainly in the news these days. The introduction of ChatGPT, an artificial intelligence system capable of writing prose to prompts input by users, captured headlines around the world.

Imagine ChatGPT as a digital oracle, a modern-day embodiment of the ancient Greek oracle at Delphi. Just as the priestess at Delphi channeled divine wisdom to answer questions, ChatGPT channels human knowledge and language to provide insights and information. At its core, ChatGPT is a neural network, a complex web of interconnected nodes modeled after the human brain's neural networks. But unlike the oracle at Delphi, ChatGPT doesn't rely on mysticism; it relies on data and mathematics.

ChatGPT's knowledge is drawn from a vast dataset of text from the internet, books, articles, and more. It learns by analyzing the patterns, context, and relationships within this vast trove of information. This allows it to grasp the nuances of language, understand context, and generate humanlike responses.

When you pose a question to ChatGPT, it's like consulting the oracle. Your query is processed through layers of neurons, where the neural network tries to understand the meaning, context, and intent behind your words. This is where the magic happens. ChatGPT sifts through its vast repository of knowledge to craft a response that's tailored to your question.

But like any oracle, ChatGPT has limitations. It doesn't possess consciousness or independent thought. Instead, it draws from its training data to provide answers. It can sometimes generate plausible-sounding responses that may not be entirely accurate or unbiased.

Furthermore, like the oracle's prophecies, ChatGPT's answers can be influenced by the way questions are phrased. Careful wording can elicit more accurate responses, just as the phrasing of questions at Delphi could shape the oracle's answers.

In essence, ChatGPT is a modern marvel that blends data, mathematics, and language to offer insights and information. It's an oracle for the

A staff member at the 2020 World Artificial Intelligence Conference in Shanghai inspects a robot at the event.

A robot called ET, powered by artificial intelligence, writes Spring Festival couplets in Hangzhou, China.

digital age, deciphering the mysteries of human knowledge and language, albeit without the mysticism of ancient Delphi.

Now here's the kicker (and here I'm getting a bit personal): The previous seven paragraphs were written by ChatGPT. To generate these paragraphs, I (Jim Trefil, the author of this part of the book) prompted it to write 300 words explaining how ChatGPT works using the writing style of James Trefil. I came up with this strategy because I thought that assessing ChatGPT's response would give us insight into AI's limitations and potential. Let's see how it did.

▶ Grading ChatGPT

FOR STARTERS, I'VE BEEN writing for almost 50 years, and I don't think I've ever referred to the Oracle of Delphi. The whole business of referring to a physical process—the ChatGPT computer program's method for writing—as magic is deeply repugnant to me. My goal in writing about any scientific subject is to uncover the basic simplicity about seemingly complex systems, be they stars or molecules. I want my readers to understand what is going on and see how the universe really works, not to stand in awe of something perceived as magical. So to use the Oracle of Delphi as a way to explain ChatGPT is a deeply flawed response to my assignment.

I don't know how ChatGPT did on copying my writing style. Scholars typically look at things like the average number of syllables per word and the number of compound sentences to judge these kinds of similarities. It must have done OK, however, because a colleague that I asked to read this chapter didn't realize ChatGPT was a ghostwriter until I told him.

As far as I am concerned, however, the fact that the AI system missed the central point of all my writing completely overshadows the question of style. I'm afraid ChatGPT and its colleagues have a long way to go before I let them write my books.

▶ Back to the Question: What Is AI?

CHATGPT ASIDE, WE STILL face large questions about what artificial intelligence is and how it is going to change our lives. As the robots Spot and Sophia showed us (see pages 222–233), AI does two things: (1) It helps computers teach other machines to accomplish tasks more effectively than humans can, and (2) it helps a computer learn and express itself like a human.

You'll notice that both these functions involve tasks we normally relate to the human brain. That's usually where AI discussions become squishy, because we really don't understand how the human brain produces intelligence, so it's difficult to compare to a computer. It might be best to adopt the tongue-in-cheek definition that computer people use: AI is anything we didn't think computers could do 10 years ago.

In fact, the AI revolution seems to have snuck up on us. Voice recognition systems like Siri, Alexa, and Google Assistant brought AI into our daily lives. The navigation system in your car or on your phone uses AI to analyze route possibilities and traffic patterns to get you from point A to point B most efficiently. Companies testing autonomous vehicles use AI to determine where a car is, what's around it, and how to get it to its destination. In the future, we expect AI to drive our cars, fight our battles, and keep our cities safe; AI is already encroaching on these tasks today.

MACHINE LEARNING

WHEN COMPUTERS WERE first introduced in the mid-20th century, most people thought they were glorified typewriters. Computers could do what they were told to do, but that was all. How things have changed! Now we have machines that can write their own programs using AI, the most common such system referred to as machine learning.

Suppose a programmer wants a computer to determine whether there is a dog in a given picture. Several things are needed to start the program. Programmers would have to supply the computer with pictures of dogs and some sort of rule (or algorithm) that tells it how to compare the images shown with pictures in its storehouse.

At this point the system has to be trained. This process involves showing the computer many pictures of dogs and letting it proceed by trial and error to find the characteristics that match the images in the pictures provided. The computer might start, for example, by guessing that dogs have four legs. If this seems to work with the pictures of dogs it has been given, it might adopt "four legs" as a criterion for defining a dog. It will then guess at another criterion—perhaps "feathers"—and apply that to its collection. If it found that feathers are not a characteristic of dogs, it should drop that requirement. As the machine continues applying new criteria, it would hopefully get closer to comprehension with each step.

This simple example illustrates two important points about AI systems: First, the system will only be able to pick out pictures of animals similar to what it has been shown. That's why negative learning is as important as positive learning—the system has to be taught that elephants are not dogs, for example, even though they have four legs. Second, to reach comprehension, all different breeds of dogs need to be represented in the input sampling. This means that developing AI systems for complex problems requires assembling huge arrays of input, or training sets. A company trying to identify possible customers, for example, might want training sets with data on millions of individuals.

▶ Learning About Learning

AS THE PROBLEMS AI DEALS with become more complex, human beings will begin to lose the ability to follow the details of how the AI system is rewriting its program. In other words, it can be hard to follow a computer's reasoning if its goals are multifaceted. This has given rise to a new academic field known as explainable AI, which is devoted to figuring out the methodology behind these complex AI programs and how they're reaching the conclusions they do.

Many fields of research are dedicated to developing new AI systems. One example of such research operates on the idea that the best way to go forward is to build computer systems that mimic the human brain. For reference, the human brain has about 86 billion neurons, each connected to as many as a few thousand other neurons. When you look around, photons trigger electrical signals in your optic nerve, which connects to a layer of neurons in the back of your skull. Depending on the details of the incoming signal, some of those neurons will fire (that is, send a signal forward to the next layer of neurons) and others will not. This process is repeated in subsequent layers, with the pattern of neurons engaged in firing becoming more complex as your brain builds up the visual image you perceive.

We are still in the process of understanding this complex neural system, but if you replace the neurons we've just described with transistors, you will have a type of AI system referred to as a neural net. Such systems are being used more and more in computers today. In fact, Sophia uses one to

The AI robot Ai-Da stands ready to appear at a House of Lords inquiry about tech and creativity in London in 2022.

interpret facial expressions. According to Hanson Robotics, Sophia's facial recognition training used a dataset of tagged photographs displaying seven expressions: happy, sad, angry, fearful, disgusted, surprised, and neutral. Now, through cameras in her eyes and chest and the trained neural net, Sophia can recognize emotions in the humans conversing with her.

▶ How ChatGPT Works

ANOTHER EXAMPLE OF A neural system is a large language model, best exemplified through what's popularly known as ChatGPT. The *P* in ChatGPT stands for pretrained, which means that it has gone through a training process like the one outlined. The *G* stands for generative, computer-speak for the fact that the program can produce a variety of outcomes, including text, pictures, sounds, and so on. And the *T* stands for transformer, the ability of the machine to convert any input into text or spoken word.

The training sets for large language models include examples of actual human speech or writing. The term "large" is truly appropriate for these training sets, which might incorporate billions of web pages for writing samples, or *trillions* of individual words.

Once the training set is established,

An art installation in Turkey uses AI to create a historical archive of Ottoman documents.

the real construction of text can proceed. Basically, the algorithm in the large language model searches through the training set to find a candidate for the next word in the sentence it is constructing. In doing so, the model has to account for all the words in the sentence, or the context, and predict the next word based on the words preceding it. This is the process by which, word by word, the system constructs its final text. If we

AI DOOMSDAY SCENARIOS

As often happens with new technologies, the advent of AI has given rise to some scary pictures of the future as well. Many scenarios involve a point of no return called the singularity: a monumental tipping point when an AI system can rewrite itself so fast that no human can control what it is doing. Then, these visionaries predict, the computer will develop a kind of consciousness and could acquire malicious, power-hungry motives with disastrous consequences for human beings.

But one amusing version of the singularity comes from the Oxford philosopher Nick Bostrom. He imagined an AI system, originally designed to make paper clips, going past the singularity and becoming a machine whose overriding goal is to convert all matter into paper clips—including humans, their homes, and basically all aspects of modern life. Even though the AI didn't have anything overtly against humans, it nonetheless developed a goal that directly harmed humans—it just needed our atoms to make paper clips.

Bostrom's story ends with the entire universe becoming one massive pile of paper clips—but wait: In a future where engineers are smart enough to create such advanced artificial intelligence, would they really neglect to build a machine without an off switch?

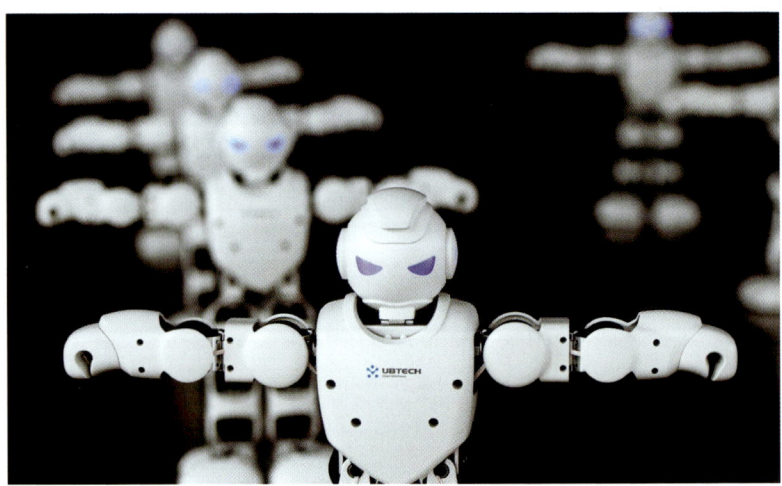

Maybe if robots are small enough, they won't be able to hurt us.

asked an AI system how to refer to a dog, the steps in the process might produce the following:

The

The dog's

The dog's name

The dog's name is

The dog's name is Rover.

This is just fine—each choice represents the most probable next word in the dataset. Suppose, however, that the second most probable choice for the third entry produced is "The dog's *breed* …" This would lead to a very different sentence. If we allow the machine to make random choices in the word-building process—by occasionally choosing the second most likely word instead of the most likely, for example—we can introduce an element of randomness into the sentence—and this can actually be a benefit. Randomness allows the machine to try many different ways of approaching a given problem to find what we might call creative solutions.

At this point, the GPT system stops looking at individual words and begins looking at examples of real sentences and paragraphs—input data supplied by human beings. Consider the following two sentences:

The ball hit the bat, and it flew away.

The bat hit the ball, and it dropped to the ground.

English speakers will recognize that "it" refers to different things in these two sentences. It refers to the ball in the first sentence, the bat in the second. At this stage of training for the AI system, the machine receives many examples like this, some actually supplied by human beings. Using the same sorts of techniques already described, the machine will develop ways of dealing with the complexities of real human speech. Its output will sound natural like Sophia's, and not machinelike. Only when it has gone through all these steps is it ready to take on the tasks for which it was developed.

THE FUTURE OF AI

RIGHT NOW, LARGE language models are mainly used for generative AI, but many developers are using AI systems to write computer code—that's a machine programming a machine. This potential could lead to a whole host of other applications in the future, but AI will have to overcome a number of obstacles to reach its full potential.

The first obstacle to improved AI systems is the sheer volume of computing power that these systems require. The cost of building and manipulating a general training set has overwhelmed many scientists. After all, a fully developed training set for a general GPT system has to include, in principle, all of human knowledge to be comprehensive. Just developing a reasonably appropriate training set for such a system is a massive barrier to entry into the AI market for new companies. (For the record, ChatGPT is produced by a company called OpenAI, which is backed by the financial might of Microsoft.)

One way around these costs is for a research company to limit the scope of (that is, specialize) the system being developed. An AI system that can help categorize and recommend movies, for example, doesn't need a training set including medical terms. Much of the current development of large language models is concentrated on these programs, often called narrow AI. A car manufacturer might develop AI systems to test cars on an assembly line, for example, while a food company might want programs to test limited-edition flavors of their products.

The second barrier to AI advancement is the need for huge amounts of data for training sets. Companies like Google train their machines on datasets of more than a trillion words, the equivalent of about 100 million 300-page books. Companies trying to break into the market don't have those resources, so they're trying alternative strategies, like signing contracts with agencies that can provide them with exclusive training sets.

The biggest concern about AI's rapid advancement, however, might be its toll on the planet. Data centers consume massive amounts of raw materials, water, and energy—a ChatGPT search uses 10 times the electricity as a Google search, for example—so the tech companies developing AI will have to innovate across all the areas we've discussed. They'll need to "green" their data centers with renewable energy, make efficient algorithms that harness less electricity than they do now, build under procedures that regulate their facilities' environmental impact, and measure that impact so companies can create new standards from the information. While it appears that humanlike AI is in our future, real humans will be central to this tech proliferating responsibly and ethically.

▶ Have We Found the Next Big Thing?

SEARCHING FOR THE FUTURE of science, innovation, and technology often brings humans into paradoxical realms of excitement and anxiety. Fusion reactors, electric aircraft, timber skyscrapers, and AI-powered robots are all exhilarating to see as they become reality, but at the same time, they all represent change, which involves an intimidating unknown. As exemplified through AI, we seem to have a habit of taking the unknown and extrapolating it into disaster: Fusion reactors could lead to explosive consequences, electric aircraft could fall from the sky, timber skyscrapers could go up in flames, and AI robots could turn on their creators. Every advance carries doomsday thinking like a barnacle.

We tend to forget that behind these innovations are brilliant, prepared, experienced scientists and engineers. They have our concerns in mind. The NIF reactor is replete with maintenance protocols and kill switches. Many eVTOLs use multiple motors so they can stay aloft should one unexpectedly fail. Ascent's wood is tested against infernal temperatures. Spot and Sophia have parameters around their training sets and learning models. Of course, some innovations will fail, or disappoint, or languish in developmental limbo—we've tapped into a few of them—but that should illuminate the success stories as incredible feats of problem-solving. That's the fundamental process of science.

As our planet battles climate change and societies debate how to address the crisis, science stands not just as a solution that can reverse negative trends, but also as an optimistic posture toward the world as a whole. Things are uncertain, yes, but the deliberate steps of assessing the problems, hypothesizing an answer, and testing new ideas creates—at the very least—empowerment. We don't know what the future will look like, but we can learn about what's possible within it. We can't snap our fingers and cool global temperatures, but we can gather data and organize responses that take us one step closer to a healthier environment. We don't know what science will reveal to us next, but we can trust the scientific method to bring us reliable, replicable results. Despite the unknown, creating the next big thing is a statement of belief in our world, and in ourselves. As this book has shown, there's a lot worth believing in.

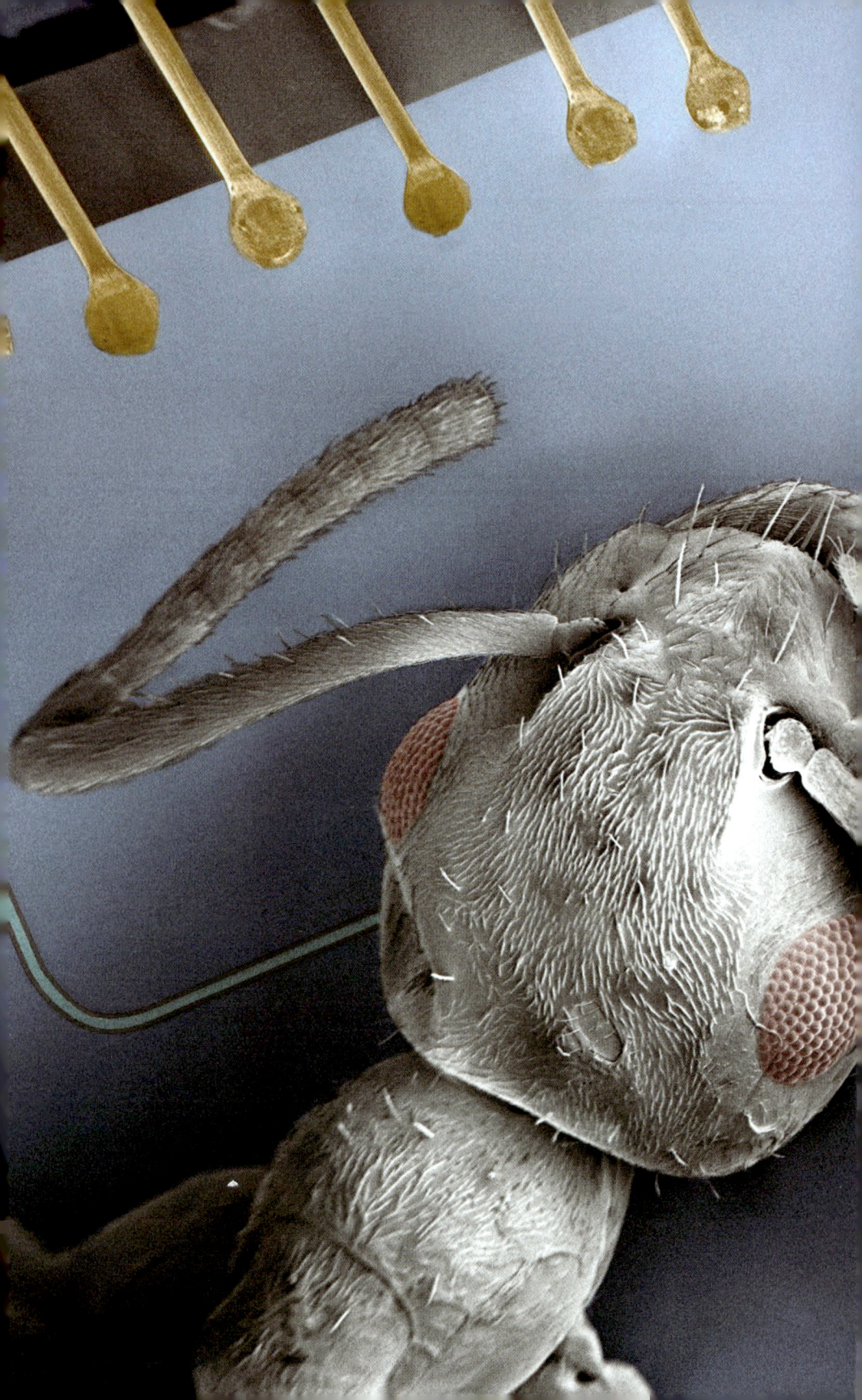

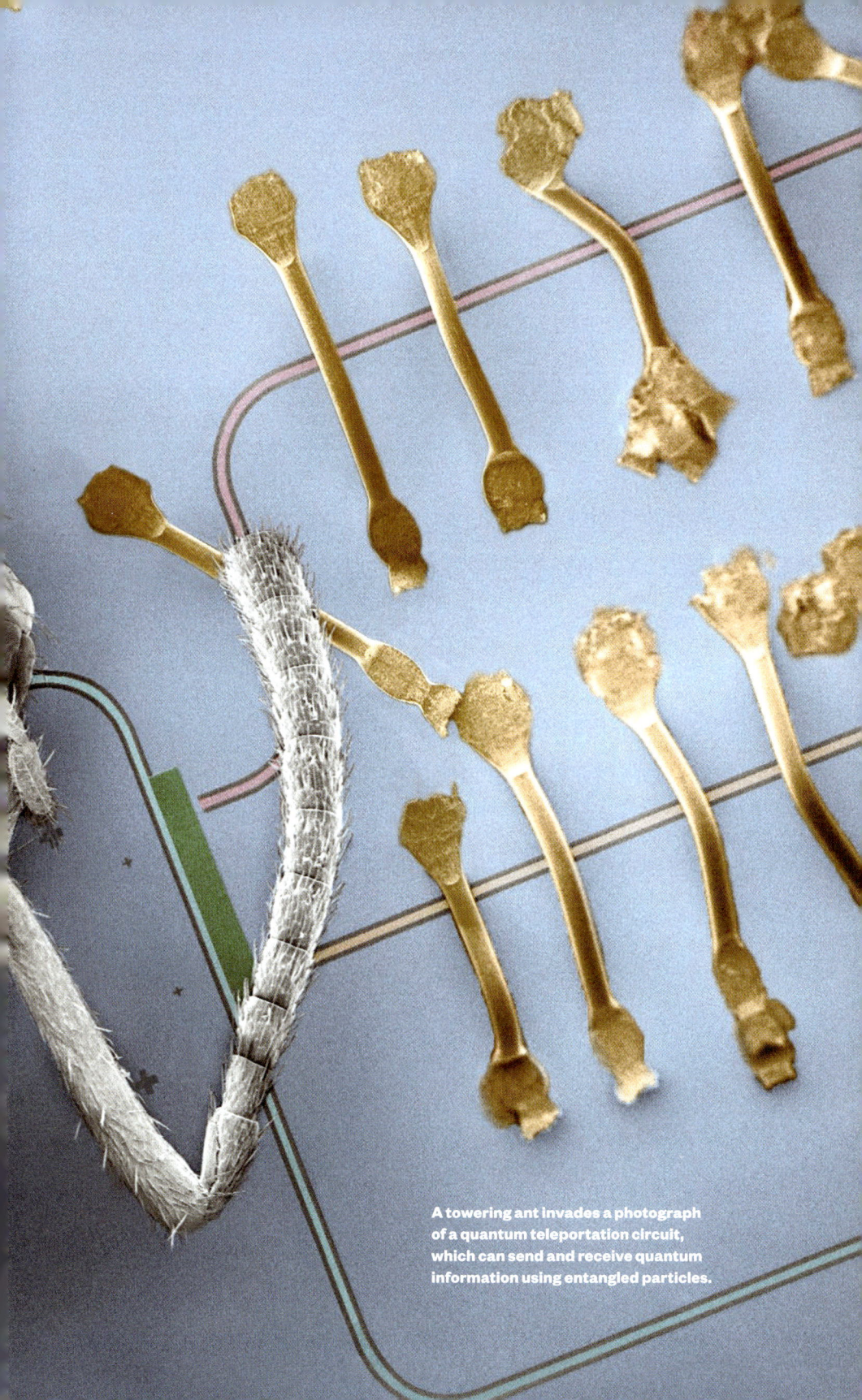

A towering ant invades a photograph of a quantum teleportation circuit, which can send and receive quantum information using entangled particles.

ACKNOWLEDGMENTS

WHEN I PERUSED THE ICONIC National Geographic Kids' learner *How Things Work,* it struck me how every page was colorful, fascinating, and easy to understand. If only all my old school textbooks had been so welcoming! So imagine my glee when National Geographic approached me to co-author a science book that peers into the future. "Honored" and "excited" don't begin to touch my emotions working on this book. I'm eternally grateful for the stewardship of senior editor Susan Hitchcock, our captain from takeoff to landing. Susan brought me on board and believed through many bumpy roads that this collaborative effort would flourish. And I can't say enough about the smarts and editing talent of associate editor Tyler Daswick. This guy can cut, polish, and primp a word mess like I've never seen. Word, my brother.

The imagery and artistic design of this book make the words leap from the page. Thanks to art director Sanaa Akkach, designer TJ Tucker, and photo editor Matt Propert for elevating *The Next Big Thing* to something worthy of display. It's beautiful. And I must thank Dr. James Trefil, the smartest guy in the room, who would never ever admit it. Professor Jim is a treasure trove of knowledge, a healthy skeptic, and a gifted teacher, and he has a heart of gold. My hosts at each on-site reporting visit for this book took several annoying follow-up calls and emails from me—thanks for your grace and patience. To my television producers and field crews who drove headfirst into the storms with me, without complaints about the danger of the conditions or my moodiness, thank you. Together we endured the elements that feed my soul. Lastly, co-authoring this book has been a challenging and rewarding journey for me, one I'm so very grateful to be able to share with you, the reader. —R.M.

ILLUSTRATIONS CREDITS

Cover, Colin Anderson/Stocksy; back cover, (LE), Brandon Moser/Alamy Stock Photo; (CT), Education Images/Universal Images Group via Getty Images; (RT), Amanda Voisard/Reuters/Redux; 2–3, Sandia National Laboratories/eyevine/Redux; 6–7, Jacek Nowak/Alamy Stock Photo; 8, Sibila and Pavel/Stocksy; 13, Planet; 14–5, Billy Hustace/Corbis Documentary/Getty Images; 16–19, Lawrence Livermore National Laboratory; 20 (UP), Damien Jemison/Lawrence Livermore National Laboratory; 20 (LO), Jason Laurea/Lawrence Livermore National Laboratory; 23–7, Lawrence Livermore National Laboratory; 28–9, Keith Ladzinski/National Geographic Image Collection; 31 (UP), rusm/E+/Getty Images; 31 (LO), Tom Hegen; 32, mikroman6/Moment/Getty Images; 34, Matthieu Paley/National Geographic Image Collection; 37, SPACEX/Science Source; 39, Isabella Moore/The New York Times/Redux; 40 (UP), Jason Edwards/National Geographic Image Collection; 40 (LO), Minoru Kuriyama/Alamy Stock Photo; 42–3, Tom Hegen; 45, Vu Khoa Nguyen Khanh/Alamy Stock Photo; 47, Sean Pavone/Alamy Stock Photo; 48, Robert Clark/National Geographic Image Collection; 50 (UP), Tom Hegen; 50 (LO), George Steinmetz/National Geographic Image Collection; 52, Paul Langrock/laif/Redux; 55, Jenn Ackerman and Tim Gruber/The New York Times/Redux; 56–7, David Guttenfelder/National Geographic Image Collection; 59, King Abdullah University of Science and Technology; 61, Nemes Laszlo/Science Source; 63, Mauro Rodrigues/Panther Media/Alamy Stock Photo; 67, Luca Locatelli; 68–9, Luca Locatelli/National Geographic Image Collection; 70–1, Philippe Psaila/Science Source; 73, Joe McNally/Getty images; 74 (BOTH), Lawrence Livermore National Laboratory; 76, Fouad A. Saad/Shutterstock; 79, chain45154/Moment/Getty Images; 81, Various images, text, or other works included in this material are copyright © 2007 or later by NuScale Power, LLC. All rights reserved. The works owned by NuScale Power, LLC may not be copied or used to create derivative works without NuScale's express permission; 84–5, Markel Redondo; 86, Archer; 89 (UP), Amanda Voisard/Reuters/Redux; 89 (LO), Archer; 90–1, John D. Ivanko/Alamy Stock Photo; 92–3, Archer; 95 (UP), Amr Alfiky/Reuters/Redux; 95 (LO), Gilles Rolle/REA/Redux; 96–7, Francisco Negroni; 99, Yankovsky88/Shutterstock; 100, RGR Collection/Alamy Stock Photo; 102, Samya Jakane/Alamy Stock Photo; 103 (UP), Education Images/UIG/Alamy Stock Photo; 103 (LO), Costfoto/NurPhoto via Getty Images; 104, Photo Researchers/Science History Images/Alamy Stock Photo; 106 (UP), Scenics & Science/Alamy Stock Photo; 106 (LO), Thomas Pflaum/VISUM/Redux; 109, Chris Ratcliffe/Bloomberg via Getty Images; 110–1, Vincent Laforet; 113 (UP), Abdullah Ali Abdulridha/National Geographic Image Collection; 113 (LO), CFOTO/Future Publishing via Getty Images; 114, Davide Monteleone/National Geographic Image Collection; 117, Edison Tech Center; 118, aslysun/Shutterstock; 120, Bernd Hartung/Agentur Focus/Redux; 121, infinity/Adobe Stock; 123, Philippe Huguen/AFP via Getty Images; 124–5, CB Studio/Stocksy; 127, Pattarawit Chompipat/Alamy Stock Photo; 128, sivvector/Adobe Stock; 129, Love Employee/iStock/Getty Images; 130 (UP), ullstein bild via Getty Images; 130 (LO), English Heritage/Heritage Images/Getty Images; 133, Maxx-Studio/Shutterstock; 134–5, Frederic J. Brown/AFP via Getty Images; 137, Felix Odell/The New York Times/Redux; 138–9, Felix Mizioznikov/Shutterstock; 141, SSPL/UIG/Bridgeman Images; 143 (UP), Jordi Busque/National

ILLUSTRATIONS CREDITS

Geographic Image Collection; 143 (LO), Michael Abid/mauritius images/Alamy Stock Photo; 144, burakyalcin/Shutterstock; 147, NASA/SDO; 150 (UP), Michael Melford/National Geographic Image Collection; 150 (LO), Jens Buettner/DPA/Alamy Stock Photo; 152-3, Sorin Colac/Alamy Stock Photo; 154-8, New Land Enterprises; 161 (BOTH), Thornton Tomasetti; 162-3, New Land Enterprises; 164-5, Li Ding/Alamy Stock Photo; 166, Josep Novellas/500px/Getty Images; 167, Matthew Micah Wright/The Image Bank/Getty Images; 171, Morris MacMatzen/Getty Images; 172, VJUS; 174, Education Images/Universal Images Group via Getty Images; 177, Connect Images/Kevin G. Smith/Design Pics; 178-9, Science Photo Library/Science Source; 181 (UP), Cédric Gerbehaye/National Geographic Image Collection; 181 (LO), weisschr/iStock/Getty Images; 183, Spencer Selvidge/Reuters/Redux; 185, Florian Neukirchen/Alamy Stock Photo; 186, Diva Amon and Craig Smith, ABYSSLINE Project; 189, AP Photo/Jon Fahey; 190 (UP), Joris van Gennip/laif/Redux; 190 (LO), John Cancalosi/Nature Picture Library/Alamy Stock Photo; 192-3, Tamara Merino/National Geographic Image Collection; 195 (UP), Luca Locatelli/National Geographic Image Collection; 195 (LO), David Guttenfelder/National Geographic Image Collection; 196, Luca Locatelli/National Geographic Image Collection; 199, Pictorial Press/Alamy Stock Photo; 200, David Liittschwager/National Geographic Image Collection; 201, VectorMine/Adobe Stock; 203, Mohamed Abdulraheem/Shutterstock; 206-7, Ina Fassbender/AFP via Getty Image; 209 (UP), Alexander Tolstykh/Shutterstock; 209 (LO), Jim West/REPORT DIGITAL-REA/Redux; 210, Jonathan Blutinger/Columbia Engineering; 213, Patrick Strattner/Guardian/eyevine/Redux; 214, Yuriy Dyachyshyn/AFP via Getty Images; 217, ESA/eyevine/Redux; 218 (UP), Joel Saget/AFP via Getty Images; 218 (LO), AP Photo/Jens Meyer; 220-1, Pete Hansen/Shutterstock; 222, Giulio Di Sturco/contrasto/Redux; 225, Jan Stradtmann/OSTKREUZ/Redux; 226-7, Boston Dynamics; 229, Hanson Robotics; 230 (UP), Giulio Di Sturco/contrasto/Redux; 230 (LO), Tyrone Siu/Reuters/Redux; 232-3, AP Photo/Bela Szandelszky; 234-5, Sean Pavone/Alamy Stock Photo; 237, NASA, ESA, CSA, STScI/Image processing: Joseph DePasquale (STScI), Alyssa Pagan (STScI), Anton M. Koekemoer (STScI); 239 (UP), ImageSource/REA/Redux; 239 (LO), Justin Mullet/Stocksy; 240, Anatolii Stoiko/Shutterstock; 241, petrroudny/Adobe Stock; 243, Lawrence Livermore National Laboratory; 244, wu kailiang/Alamy Stock Photo; 245, Demirkan/Adobe Stock; 247, Christie Hem Klok/The New York Times/Redux; 248-9, TEK IMAGE/Science Source; 250 (UP), Babak Tafreshi/National Geographic Image Collection; 250 (LO), Peter Horree/Alamy Stock Photo; 253, Bettmann/Getty Images; 254, Michael Stuparyk/Toronto Star via Getty Images; 255, PVDE/Bridgeman Images; 257, dencg/Shutterstock; 261, Brandon Moser/Alamy Stock Photo; 262-3, Mattia Balsamini/contrasto/Redux; 265, Jason Koxvold, Wired, © Condé Nast; 266, Isabella Moore/The New York Times/Redux; 268-9, Mattia Balsamini/contrasto/Redux; 270, Gerville/iStock/Getty Images; 273, Robert Clark/National Geographic Image Collection; 276-7, Max Aguilera-Hellweg/National Geographic Image Collection; 279 (UP), Aly Song/Reuters/Redux; 279 (LO), Visual China Group via Getty Images; 283, Elliott Franks/eyevine/Redux; 284-5, Chris McGrath/Getty Images; 286, FeatureChina via AP Images; 290-1, Quantum Device Lab.

INDEX

INDEX

INDEX

INDEX

ABOUT THE AUTHORS

▶ Rob Marciano

Rob Marciano is the senior meteorologist and national weather correspondent for CBS News. His career in weather spans 30 years of reporting from hundreds of storms, disasters, and natural phenomena, including blizzards, hurricanes, tornadoes, wildfires, eclipses, and volcanic eruptions. Marciano graduated from Cornell University with a bachelor's degree in atmospheric science and holds the American Meteorological Society Seal of Approval as a Certified Broadcast Meteorologist. When not forecasting weather or investigating the wonders of science, Marciano enjoys the outdoors and spending time with his two children, Madelynn and Mason. He lives in Rye, New York.

▶ James Trefil

James Trefil is the Clarence J. Robinson Professor of Physics at George Mason University. He held positions at the Stanford Linear Accelerator Center, European Organization for Nuclear Research (CERN), Laboratory for Nuclear Science at MIT, German Electron Synchrotron laboratory in Hamburg, Fermi National Accelerator Laboratory, Argonne National Laboratory, and the Universities of Illinois and Virginia before joining the George Mason faculty in 1987. He is the author of more than 50 books, including National Geographic's *Space Atlas*, and is internationally respected for his ability to explain complex science to the general public. He lives with his wife, Wanda, in Fairfax, Virginia.

Since 1888, the National Geographic Society has funded more than 15,000 research, conservation, education, technology, and storytelling projects around the world. National Geographic Partners distributes a portion of the funds it receives from your purchase to National Geographic Society to support their mission to illuminate and protect the wonder of our world.

National Geographic Partners, LLC
1145 17th Street NW
Washington, DC 20036-4688 USA

Get closer to National Geographic Explorers and photographers, and connect with our global community. Join us today at nationalgeographic.org/joinus

For rights or permissions inquiries, please contact National Geographic Books Subsidiary Rights: bookrights@natgeo.com

ISBN: 978-1-4262-2329-7

The authorized representative in the EU for product safety and compliance is Disney Trading B.V., Asterweg 15S, 1031 HL, Amsterdam, The Netherlands email: DCP.DL-EU.bookscontact@disney.com

Printed in Malaysia

26/IVM/1

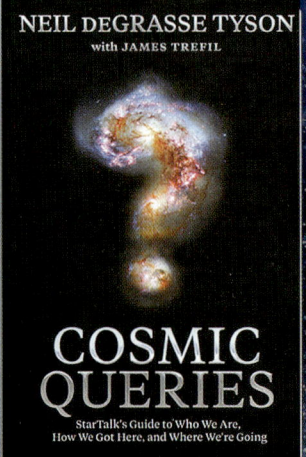
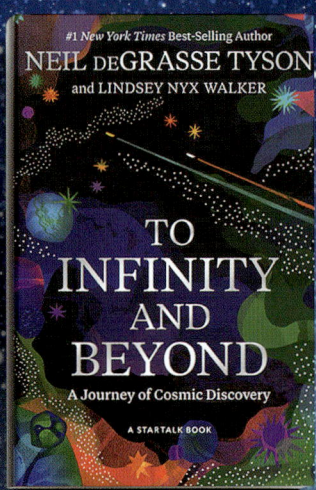
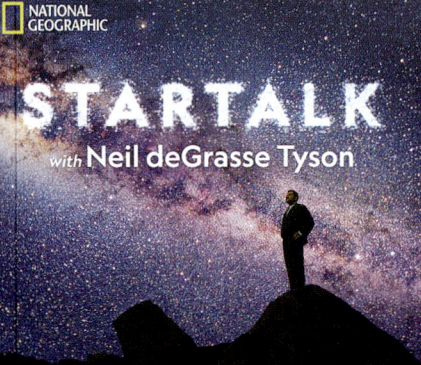